W9-CPN-418

The Sun Kings

The Sun Kings

The Unexpected Tragedy of
Richard Carrington and the Tale of How
Modern Astronomy Began

STUART CLARK

Princeton University Press
Princeton & Oxford

Published by Princeton University Press, 41 William Street,
Princeton, New Jersey 08540
In the United Kingdom: Princeton University Press, 3 Market Place,
Woodstock, Oxfordshire OX20 1SY

ISBN-13: 978-0-691-12660-9
ISBN-10: 0-691-12660-7

Library of Congress Control Number: 2006940123

British Library Cataloging-in-Publication Data is available

This book has been composed in Adobe Caslon

Printed on acid-free paper. ∞

press.princeton.edu

Printed in the United States of America

1 3 5 7 9 10 8 6 4 2

To Nikki,

WHO TEMPORARILY LOST HER HUSBAND TO
THE COURT OF THE SUN KINGS.
I'M BACK NOW.

Contents

List of Illustrations ix
Acknowledgments xi
PROLOGUE The Dog Years 1
ONE The First Swallow of Summer 9
TWO Herschel's Grand Absurdity 25
THREE The Magnetic Crusade 47
FOUR The Solar Lockstep 58
FIVE The Day and Night Observatory 71
SIX The Perfect Solar Storm 80
SEVEN In the Grip of the Sun 93
EIGHT The Greatest Prize of All 98
NINE Death at the Devil's Jumps 117
TEN The Sun's Librarian 129
ELEVEN New Flare, New Storm, New Understanding 148
TWELVE The Waiting Game 168
THIRTEEN The Cloud Chamber 179
EPILOGUE Magnetar Spring 188
Bibliography 191
Index 207

Illustrations

The sunspot groups responsible for the 2003 Halloween flares 5
Richard Carrington's sketch of the solar flare 14
William Herschel 45
John Herschel's 20-foot telescope at Feldhausen, South Africa 55
John Herschel 59
Richard Carrington's house and observatory at Redhill, Surrey 72
An engraving of the aurora 85
Warren de la Rue and his team at Rivabellosa 106
Warren de la Rue's picture of the totally eclipsed Sun 108
George Biddell Airy 132
E. Walter Maunder 141
The Royal Observatory at Greenwich 144
George Ellery Hale's July 1892 solar flare photograph 149
Annie Maunder's photograph of the 1898 eclipse 157
An accompanying sketch 157

Note on Illustrations

No portrait of Richard Carrington is known to exist. He signed a letter to George Airy, along with nine other scientists, urging the Astronomer Royal to have a portrait taken. In the letter, all ten gentlemen claimed to have submitted themselves to the camera of Messrs Maull and Polybank (55 Gravechurch Street, London). The search continues. . . .

Acknowledgments

Peter Tallack for believing in the project and never giving up on it.

Ingrid Gnerlich for just "getting it," right from the moment she saw the proposal.

Nicola Clark for sharing the ups and downs of this project with me, and being my essential sounding board in all matters of the English language.

Peter Hingley for being unfailingly supportive; without him this book would have been a lesser work.

Mary Chibnall for always succeeding in fulfilling my library requests, no matter how vague my references.

Norman Lindop for sparking my interest in Carrington with a thesis that collected so much of the Carrington material together.

Norman Keer, one of the last inhabitants of Carrington's Redhill observatory home, for sharing his decades of amateur sleuthing and his unalloyed passion for all things Carrington.

Owen Gingerich for taking my breath away with his enthusiasm for the book.

Dava Sobel for being so interested in the project.

Latha Menon for making the initial running.

Sheila Simons for helping me investigate the Carrington family's births, deaths, and marriages.

All those who helped with technical and historical advice, read parts or all of the manuscript, and provided feedback: Pål Brekke, Lars Bruzelius, Ed Cliver, Bernhard Fleck, Anthony Kinder, Chris Kitchin,

Nick Kollerstrom, Gurbax Lakhina, Jack Meadows, Paul Parsons, Jay Pasachoff, Peggy Shea, George Siscoe, Harlan Spence, Willie Soon, Amarendra (Bob) Swarup, Bruce Tsurutani, David Whitehouse, and David Willis.

To everyone who helped and to all those whose papers I read: my thanks are beyond measure. You helped bring these deceased astronomers back to life for me and, by extension, for the readers too. Any success this book enjoys reflects on you.

PROLOGUE

The Dog Years

They say that dogs age seven times faster than humans. No one is more aware of this than the men and women in charge of an aged electronic watchdog that fights a daily battle against decrepitude in the name of science. The Solar and Heliospheric Observatory, known universally as SOHO, is an electronic beast stationed 1.5 million miles away, in one of the most hostile environments any spacecraft has ever been expected to inhabit. Here, SOHO is perpetually bombarded, not just by light, heat, and X-rays from the Sun, but also by a wind of smashed atoms hurled outwards by the unpredictable solar forces.

Had this watchdog been a flesh-and-blood animal, the onslaught would long since have triggered the deathly creep of cancer. In the machine world, the equivalent is an inexorable degeneration as the subatomic bombardment gradually eats away at the spacecraft's electronic organs. By 2003, after nearly eight years in space, SOHO had lost the use of certain cameras and other electronic systems. Its antenna would not point straight anymore, and its ability to harvest sunlight for power was down nearly a fifth. Yet it soldiered on, constantly monitoring the Sun's boiling surface for clues that might one day solve a century-and-a-half-old mystery: why enormous explosions occasionally plague the Sun. And more importantly, how they affect us when the Earth accidentally gets in the way of the blast.

The Sun is the heart of our solar system. It is an enormous sphere of gas, over a hundred times the diameter of the Earth. Its surface temperature is 6,000 degrees Celsius; its center is at well over 10 million degrees. Its gravity guides Earth and the other planets through their orbits; its warmth provides the lifeblood of energy for plants and animals on Earth. Also like a heart, the Sun pulsates. This is not a visible movement but rather a gradual buildup in strength and subsequent weakening of the giant magnetic bubble that emanates from within the Sun and surrounds all the planets. As befits a celestial body of some 4.6 billion years in age, each one of these magnetic heartbeats takes a leisurely eleven years, or thereabouts, to complete.

So, in the average career of a scientist, he or she can expect to see this happen four times. This makes understanding the Sun as difficult as a biologist trying to deduce the life cycle of an unknown creature by observing it just long enough to witness four beats of its heart. As a result, solar astronomy is a multigenerational science. Each new cohort works to build a finer legacy of observations for those yet to come.

No one knows when that body of evidence will be rich enough to provide the necessary insight, or when technology will be mature enough to provide the final incisive observation. Each new generation of astronomers works with the same ambition that drove their scientific forebears: that they might be the ones who finally understand the Sun. When the Sun entered an intense bout of activity in 2003, SOHO's astronomers realized they had been given the chance of a lifetime—if their spacecraft could survive.

During October and November that year, the Sun was wracked by a succession of explosions known as solar flares, the most powerful events that can take place in the solar system. A solar flare dwarfs the power of an atomic bomb, and during the fourteen days spanning the Halloween period, some seventeen of them erupted across the Sun. Each one triggered a powerful "sunquake" and smothered SOHO with a burst of debilitating radiation. Some of them also triggered major eruptions that each spewed billions of tons of electrified gas into space, pummeling anything that got in the way, be it the tiny SOHO spacecraft or the entire planet Earth.

Scientists looked on, suffused with a cocktail of excitement, awe, and dread. No one knew how long SOHO could survive under normal circumstances; under these conditions it was anyone's guess. SOHO's masters could do little but hope for the best as they sat in their offices at NASA's Goddard Space Flight Center (GSFC) in Greenbelt and watched their spacecraft receive the worst punishment of its life.

Just a few weeks before, there had been no hint of such activity on the Sun's boiling surface. In fact, it was so quiet that scientists were beginning to think that it had settled into one of its periodic dormant phases. Then the Sun began to quake.

SOHO picked up this solar heart murmur in early October and scientists began searching for its cause. They could find nothing on the visible face and so concluded that something on the far side was hurling shock waves right around the Sun. They had no choice but to wait for the Sun's leisurely rotation to bring whatever it was into view.

On 18 October they spotted a darkened patch near the eastern edge of the Sun. It was barely visible at first, little more than a small blemish. Twenty-four hours later, it had swollen into an ugly bruise, seven times larger than the Earth. It was a giant sunspot. Sunspots appear from time to time, although they are usually much smaller than this monster. They are caused when knots of magnetism burst from the Sun's interior, cooling the surrounding gas and making it appear dark in comparison to the rest of the solar surface. Oriental astronomers made the first observations of sunspots thousands of years ago, spotting them with the naked eye when the Sun passed behind light cloud or fog banks.

Astronomers now know that flares often explode above sunspots, and it was not long before this particular sunspot let rip. The first Halloween flare took place above the engorging spot on 19 October. Its blast of radiation almost immediately blacked out radio communications for about an hour on the sunlit side of the Earth. Undiminished by the outburst, the sunspot continued to grow and the Sun continued to quake. Therein lay a puzzle. This particular sunspot had been almost negligible when the scientists had first seen it, yet the Sun had been quaking long beforehand. Could that mean another, already fully formed sunspot was on its way?

Suspicions were confirmed on 21 October when SOHO transmitted the next in its never-ending sequence of images that updates every fifteen minutes. To one side of the Sun, scientists could see the aftermath of a large eruption that had taken place just beyond view, over the Sun's eastern horizon. The eruption took the form of an expanding cloud of hot gas heading off into space. Subsequent images that day revealed a second outburst of seething gas from the same location. There had to be another huge sunspot edging round from the far side. The scientists estimated that the Sun's rotation would drag it into view within the next few days.

In the meantime, they still had the first giant to watch. On 22 October, it flared once again, and this time the explosion triggered its own eruption of solar gas. Larger than a planet, the gaseous eruption contained a hellish cocktail of particles, most of them electrically charged and all of them at a few million degrees Celsius, about ten thousand times hotter than the air in a kitchen oven. As scientists watched the expanding cloud of gas head off into space, they realized that some of it would touch the Earth.

Whereas light and X-rays from a flare cross the 93 million miles separating us from the Sun in just 8 minutes, each lumbering eruption of particles takes between eighteen and forty-eight hours to arrive. As the time of impact approached, astronauts Michael Foale and Alexander Kaleri hunkered down in the International Space Station's most heavily shielded module to escape the deadly storm. Airlines instructed their pilots to reduce altitude in the hope that the Earth's atmosphere would protect the passengers and crew from higher than usual doses of radiation. They also directed flights away from polar routes, which research suggests are the most vulnerable to high radiation doses during solar storms.

About half an hour before the storm struck Earth, it swept over SOHO, blinding the cameras and building up electrical charges that threatened to short-circuit the sensitive equipment. SOHO survived, but not every satellite was so fortunate. The first electronic casualty was the Japanese Space Agency's Midori 2 weather satellite, which fell silent during the bombardment and has not been heard from since. Other satellites temporarily malfunctioned or shut themselves down to await resuscitating messages from ground controllers.[1]

At the surface of the Earth, there were few reported problems although sky watchers noticed auroras glowing in the sky. These natural

[1] Often, this was because the solar storm temporarily blinded spacecraft navigation devices. These little cameras, called *star trackers*, watch the stars, allowing a spacecraft to know which way it is pointing. Cut off the star tracker and the spacecraft has no way of knowing which way is up. To protect it from firing thrusters in all directions to try and correct its perceived balance problems, the spacecraft renders itself unconscious and waits for a wake-up call from Earth when the danger has passed.

The two sunspot groups responsible for the Halloween flares of 2003. Each group is about ten times the diameter of the Earth. (Image: NSO/AURA/NSF/Bill Livingston)

light shows are caused by solar particles colliding with the molecules in our atmosphere. They usually take place close to the Earth's north and south poles, and their intensity is recognized as a barometer for measuring the activity of the Sun. During Halloween 2003, the phantom glow of the auroras lit the sky many times.

As the Sun rotated, the sunspot continued to pump out volley after volley of electrified matter. With each shot it came progressively closer to striking the Earth a direct blow. By 26 October, the sunspot had grown to more than ten times the diameter of the Earth, making it the largest for over a decade—and it wasn't alone anymore.

The second sunspot had finally slid into view around the Sun's eastern edge, and it more than equaled the first in size. To see one giant sunspot was awesome, to see two was frightening. To herald its arrival, the second twisted knot of magnetism let loose a massive flare, blacking out some radios. Not to be outdone, the original sunspot erupted as well.

And so it continued. Each new day brought a new flare and eruption. It was no longer a case of whether the Earth would be hit, merely how strong the blast would be.

On 28 October, scientists' worst fears were realized. As the original sunspot drew in line with the Earth, it exploded with the most powerful flare yet. Fifty billion times the energy of an atomic bomb was released, resulting in almost immediate communications failures across the world. The worldwide marine emergency call system became inoperable for forty minutes, contact was lost with expeditions on Mount Everest, and faltering radios hindered crews fighting forest fires in California. Ten times farther into space than the Earth, NASA's Cassini spacecraft was orbiting the ringed planet Saturn. It, too, received a blast of radio waves, let loose by the flare.

Not only this but the flare triggered an enormous solar eruption, sending a billion tons of million-degree gas careening into space, straight at SOHO and the Earth. This was too much even for the data-hungry scientists. They commanded SOHO to switch to a low-power "safe mode," turning off the vulnerable equipment. To continue operations in the face of this new eruption would be the scientific equivalent of flying a kite in a thunderstorm with piano wire instead of string to keep control. So they closed the spacecraft's eyes and concentrated instead on simply keeping it alive.

When the storm arrived at Earth, it was ferocious. The solar flare had rocketed the eruption into space at an amazing 2,300 km/sec. (1,440 miles/sec.) As a result, the electrified gases took just twelve minutes to collide with Earth after crashing past SOHO.

Again, Earth-orbiting satellites began to behave erratically. Airlines hastily rerouted flights, instructing all aircraft to drop below the line of latitude that runs from northern Scotland, across Hudson Bay to the lower tip of Alaska, and through Russia (57 degrees north). As air traffic controllers imposed these restrictions, delays began to mount up at airports. Flying altitudes were lowered to below 25,000 feet and the additional fuel needed to plough through the thicker atmosphere soon ran up a price tag of millions of dollars.

As the particles battered the Earth's natural cloak of magnetism, erratic currents surged along northern power lines, eventually damaging power stations and blacking out fifty thousand people in Sweden. In the United States of America, power was reduced at two New Jersey nuclear units for fear they would be damaged by the power surges. Magnetic compasses swung wildly back and forth as the electrified gases that had once been part of the Sun now assaulted our planet.

As the storm abated, the sunspot fired another similarly sized barrage at Earth. Indeed, as October became November, flares and eruptions repeatedly disrupted the Earth. During this time, radio communications simply could not be relied upon, satellite television reception became patchy, mobile phones ceased to work in some countries, and the Global Positioning System (GPS) gave inaccurate readings. It was the stuff of high-drama techno-thrillers, and, as word spread across the Goddard Space Flight Center, unrelated workers would daily pop into the SOHO offices to check on the progress, both of the spacecraft and the terrible assault on Earth.

Eventually, the situation began to quieten down as the first sunspot disappeared from view over the western edge of the Sun, leaving only the second spot in view. It was about now that cameraman Ed Harriman framed a picture of the setting Sun over war-torn Baghdad as part of a documentary about the eight-month-old war. His picture caught the Sun setting behind the palls of smoke and pollution drifting across the defeated city. When he played the tape back he saw something on the face of the Sun that he had failed to notice at the time. It was the second monstrous sunspot, clearly visible on the face of the Sun. SOHO continued to watch the remaining sunspot too, as it slid across the face of the Sun and headed back to the far side. But one big surprise was still in store.

On 4 November, spacecraft again saw a solar flare erupt from above this spot and throw a massive tract of solar material into space. The X-ray monitors on several spacecraft rose until eventually they overloaded. Although they could not put an immediate figure on the outburst, the waiting scientists were certain of one thing: this was the most powerful solar flare in the cycle yet, possibly the most powerful ever recorded. As they worked with the data collected before the instruments saturated, the numbers seemed too wild. After double and even triple checks, however, there was no escape from the fact that this flare was at least twice as powerful as the one that had caused such havoc the week before.

Astronomers tracked the eruption and held their breath. If it were to strike the Earth, untold damage could be done to satellites, power stations, and other forms of technology. Radiation levels inside high-altitude airliners could reach extreme levels.

Thankfully, taking place on the Sun's horizon, the explosion was not directed toward Earth, and the eruption hurtled into deep space. Earth was caught only in the side wash, with relatively minor disruption.

There could be no complacency about this good fortune. No one had done anything clever or heroic; it had simply been a lucky escape. In the ensuing weeks and months many wondered what would have happened if such a massive solar storm had directed its full force at Earth.

The answers lie buried in the historical records of around 150 years ago. . . .

ONE

The First Swallow of Summer, 1859

I boarded the king's ship; now in the beak,
Now in the waist, the deck, in every cabin,
I flamed amazement; sometime I'd divide
And burn in many places; on the topmast
The yards and bowsprit, would I flame distinctly
Then meet and join.
—*The Tempest*, William Shakespeare

There may not have been any boldness in the design of the three-mast clipper ship *Southern Cross*, but it possessed an exquisite finesse as it slipped into the cold Atlantic waters from E. and H. O. Briggs's Boston shipyard in 1851. In the quest for speed, new clipper ships had been emerging from their berths with ever longer and sharper lines. The *Southern Cross*, however, was a throwback: shorter, at 170 feet, and more rounded than had become the norm. At the prow, a gilded eagle in full flight guided its way.

The ship was named after the beautiful constellation that sits deep in the southern sky. From Boston, these stars are never visible but all anticipated that the ship would see the celestial grouping many times as it sailed around the peak of South America on its way to and from the gold rush in California.

A writer from the *Boston Daily Atlas* was certainly convinced of the ship's quality. Following the launch, he reported in the newspaper's 5 May edition that "no doubt need be entertained of her success as a swift sailor, and what is more, of being a trustworthy vessel in a heavy sea." Eight years later, on 2 September 1859, those attributes were severely tested. The *Southern Cross* was eighty-four days out of Boston, heading for San Francisco, when Captain Benjamin Perkins Howe and the ship's company sailed into a living hell.

It was 1:30 in the morning and they were in the Pacific off the coast of Chile, fighting a tremendous gale that had been raging all night. Hailstones from above and waves from all around whipped the deck. When the wind-lashed ocean spray fell away to leeward, the men noticed they were sailing in an ocean of blood. Everywhere they looked, the pitching seascape had turned deepest red. Lifting their eyes skyward, they saw the reason. It was obvious, even through the clouds: the heavens were wreathed in an all-encompassing red glow.

The sailors recognized the lights at once. They were the southern aurora, an unexplained phenomenon whose eerie glow usually graced the skies near the Antarctic Circle, just as their northern counterparts clung to the Arctic. To see the southern lights from as far north as temperate-latitude Pacific waters was highly unusual, especially considering the intensity of the display. It should have been a treat, but maintaining their faltering control over the ship robbed the crew of the opportunity to appreciate the spectacle.

As the howling maelstrom intensified, they noticed other strange lights, much closer than those of the aurora. They were clinging to the ship itself, creating haloes around the silhouettes of the mastheads and yardarms. These new wraiths were also familiar and just as inexplicable as the aurora. Sailors knew them as Saint Elmo's fire. Their usually blue-white light often accompanied ships during extreme thunderstorms, but on this night their pallid glow had been stained the same roseate hue as the heavens above.

Sailors had named Saint Elmo's fire after their patron saint, Erasmus, who had been gored with a glowing red-hot iron hook during his martyrdom. Although the electrical discharges seen by sailors were usually a different color, their appearance in stormy weather was taken to mean their ship had come under Saint Elmo's protection. A wider audience had been introduced to the phenomenon thanks to Shakespeare's *The Tempest*, in which he cast the sprite Ariel in the role and had him describe his antics.

As the night edged toward dawn, the sailors might easily have felt the need for supernatural reassurance. During a temporary lull in the storm, they witnessed an even more astonishing display. Fiery lights loomed against the horizon as if some terrible conflagration had engulfed the Earth. At other times, vivid bolts flew across the sky in spiral streaks, heading for the zenith before exploding in silent brilliance, as if the very souls of all humanity were fleeing whatever cataclysm had smothered the planet.

The storm finally abated at dawn, and the attendant sunlight drove the aurora from the sky. Upon their arrival, on 22 October, in San Francisco, Captain Howe and the ship's officers swore that they had never witnessed anything equaling the magnificence of the 2 September auroral displays. They discovered that theirs was not an isolated experience. Most of the world had been in the electrifying grip of the auroras and, spared the *Southern Cross*'s gale, had witnessed it as a silent, skywide manifestation that inspired both awe and terror in equal measure. No one could remember seeing anything on this scale before, nor could any similarly widespread occurrence be found in the history books. The Earth had experienced something unique. But what?

The answer lay half a world away, where a wealthy Victorian gentleman, who loved nothing more than indulging his overwhelming passion for astronomy, was wrestling with his own scientific conundrum. He had been in the right place at the right time and had seen something unprecedented. Now he was trying to make sense of it.

At 33, Richard Christopher Carrington was already an accomplished young astronomer. He possessed a first-class education from Trinity College, Cambridge; had compiled a much-needed star catalog that drew praise from all; and worked as a tireless, unpaid ambassador for the Royal Astronomical Society (RAS). The only thing that eluded him was a visitation from the hand of fate, in the guise of a unique scientific discovery. Even in the modern era, such serendipity has the power to transform a great scientist into a guru. On the morning of Thursday, 1 September 1859, the day before the *Southern Cross*'s auroral sighting, fortune finally favored him with such a prize.

He was working in his grandly appointed private observatory at Redhill, Surrey. At the sight of that morning's clear sky he had hurried into the dome, cranked up the shutter, and prepared the beautiful two-meter-long brass telescope for action. He had followed the same routine since 1853, when he had resolved to make a long-term study of the Sun and the transitory sunspots that speckled its surface.

Maneuvering a board painted with distemper into position, he aligned the telescope so that it threw the Sun's image onto the straw-colored screen. Then, poking the front end of the telescope through a made-to-measure hole, he slotted a larger board into position around the telescope itself. This cast a shadow across the board, allowing him to see the Sun's eleven-inch-wide image more clearly. Two gold wires, beaten into slivers and strung inside the telescope's eyepiece, cast a diagonal crosswire on the image. Using the lines as position guides, Carrington set about sketching the entire face of the Sun, employing his enviable draftsman's skill to produce a lasting document of the exquisite details of the solar surface.

Such labors were a welcome respite from running the family brewery, a job he resented bitterly, having been forced into it the previous year by the unexpected death of his father. Where astronomy had once been his self-financed profession, now it was more of a therapy to counteract his growing frustration with the rigors of commerce.

As far as solar astronomy was concerned, today was special because an enormous sunspot complex was visible. No one knew what these blotches were. Some thought them openings in the bright clouds of the Sun, through which the Sun's true surface could be glimpsed. Others believed them to be mountaintops occasionally revealed by the Sun's shifting atmosphere. The one that Carrington gazed upon that day was huge beyond imagination. From tip to tip it was almost ten times the diameter of the Earth. Yet on the Sun, it barely stretched a tenth of the way across the fiery disk.

By eighteen minutes past eleven, he had finished the drawings and was now listening to the tick of the chronometer, recording the precise moments at which the various sunspots slipped beneath the crosswire. He would later use the timings to perform some elaborate mathematics in order to calculate the sunspots' exact positions.

Without warning, two beads of searing white light, bright as forked lightning but rounded rather than jagged and persistent instead of fleeting, appeared over the monstrous sunspot group. Momentarily taken by surprise, Carrington assumed that a ray of sunlight had found its way through the shadow-screen attached to the telescope. He reached out and jiggled the instrument, expecting the errant ray to zip wildly across the image. Instead, it stayed doggedly fixed in its position on the sunspot group. Whatever this was, it was not some stray reflection; it was coming from the Sun itself. As he stared, dumbfounded, the two spots of light intensified and became kidney shaped.

Carrington noted later that he became rather "flurried by the surprise" of being "an unprepared witness" to the event. Nevertheless, his scientific training switched back on instantly, and he hurriedly noted down the time. Then, realizing the rarity of the situation—certainly no one had ever publicly described the Sun behaving like this before—he hastened to find a witness.

Upon his return, not sixty seconds later, his excitement turned to mortification as he saw that the strange lights straddling the sunspot were already greatly enfeebled. Nevertheless, they were still visible, and he watched them drift across the giant spot. As they did so, they contracted into mere points and then abruptly vanished.

He noted the time again, 11 h 23 m Greenwich Mean Time, and sketched the position of the lights' appearance and disappearance. Then, staggered by what he had seen, he rooted himself to the telescope for over an hour, hardly daring to move, in case the mysterious flares reappeared.

His vigil was to no avail; the Sun had instantly returned to normal. In fact, he could see no indication that the strange phenomenon had ever taken place. The Sun's surface surrounding the spot and the details of the spot itself remained exactly as they had before the phantoms appeared.

Later, Carrington set to work on the mathematics. The lights had lasted but five minutes, yet in that time he discovered they had traversed 35,000 miles (nearly four-and-a-half times the diameter of the Earth). To do that, the disturbance must have moved at around 420,000 miles per hour. Such a staggering speed must have strained even Carrington's belief, because as a people the Victorians were still getting used to steam trains chuffing their way to fifty miles per hour. And that was not the end of the big numbers. Judging by the extent of the flare on his sketch, the original fireballs had each been about the size of the Earth.

Carrington would have known that such a momentous observation demanded independent confirmation if his peers were truly to believe

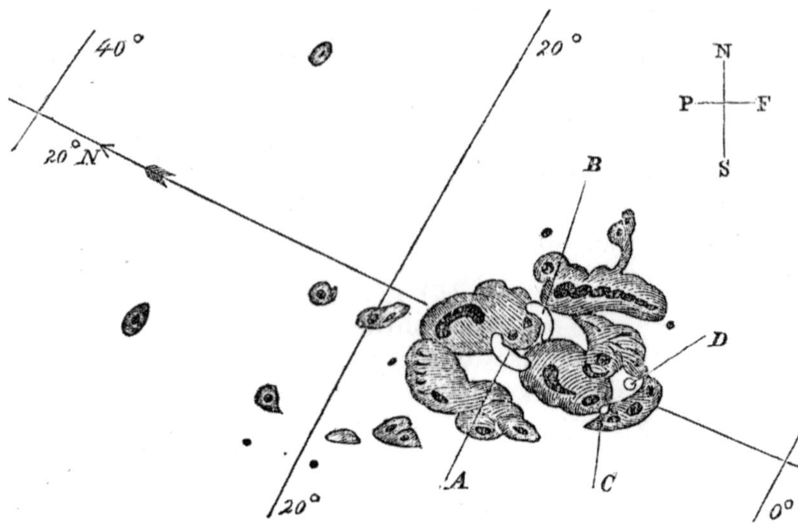

Richard Carrington's sketch of the solar flare taking place above the sunspot group on 1 September 1859. "A" and "B" represent the positions in which the two kidney-shaped spots of light appeared. "C" and "D" show the place where the shrunken lights disappeared. (Image: Royal Astronomical Society)

him. The surviving accounts do not make it clear whether Carrington succeeded in finding a witness from within his own household (he simply stated that he went looking for one, returning within a single minute). It is clear, however, that even if he had dragged out some unsuspecting soul to vouch for him, he also needed a wholly independent scientific corroboration of his sighting. It soon occurred to him that the perfect place would be the Kew Observatory, where his friend and RAS colleague, Warren de la Rue, was engaged in an experimental project to photograph the Sun every clear day. Carrington would need to visit Kew as soon as possible.

That night, about eighteen hours after Carrington's observation, as night gave way to another day and dawn crept over Europe, the Earth's atmosphere erupted with auroras. Such displays manifest themselves in various forms, each of which nowadays is denoted by a specific term. Proceeding in order of increasing magnificence and sky coverage, the first hints of a display can often be a mild *glow*, hugging the horizon. Bright *patches* (sometimes called *surfaces* by scientists) can also appear;

these sit like luminous clouds in the sky. Next is an *arc*, which stretches like a fluorescent basket handle across the sky. *Rays* are sure signs of strengthening auroral activity; they are often born from arcs and reach upward like ragged picket fence posts. Undoubtedly, the crowning glory of an auroral display is when the entire sky is wreathed in celestial fire. Then, a structure called a *corona* appears in which rays from all around the sky converge to a point. It is the hallmark of exceptional activity and a rare sight outside the polar latitudes.

Auroras come in different colors. Reddish pink and greenish yellow light comes from interactions involving oxygen atoms. Generally, the red hues come from interactions that take place at higher altitudes than those that give rise to the green tones. Purple and violet emissions, sometimes described merely as blue, derive from the atmosphere's nitrogen atoms.[1]

As the auroras of 2 September 1859 smothered the planet, those on the clipper ship *Southern Cross* were among the first, but by no means the only, people to witness their light. Accounts from mainland Chile back up the sailors' extraordinary tale. In Concepcion, located at 36 degrees 46 minutes south, almost one thousand miles closer to the equator, the aurora burst forth in the early hours of the morning. It was described as looking like "a cloud of fire, or a large *ignis fatuus*" that drifted from east to west. *Ignis fatuus* is the name given to "fool's fire," an atmospheric phenomenon similar to Saint Elmo's fire. Often referred to as will-o'-the-wisp, *ignis fatuus* is a sure sign of electricity in the air. The glow is, in fact, a natural form of the fluorescent light tube. At Santiago (33 degrees 28 minutes south), the population was treated to a brilliant display of blue, red, and yellow colors that illuminated the city for about three hours.

As these southern lights crept up toward the tropic of Capricorn, located at 23 degrees south, the northern auroras spread downward. A "perfect dome of alternate red and green streamers" sat over Newburyport, Massachusetts (42 degrees 48 minutes). It provided light "so great that ordinary print could be read as easily as in daytime." The ability to read by the auroral light was cited as a mark of its intensity in a number

[1] In the following descriptions of the auroras, I have used faithfully the colors quoted by the many and varied eyewitnesses. In general, it is easy to see which of the red, green, and blue categories they fall into.

of places. In nearby Lunenburg, Professor William B. Rogers, founder of the Massachusetts Institute of Technology (MIT), provided a "textbook" account, almost as vivid as the northern lights themselves:

> Sept. 2d, a clear sunset was followed by a peculiar greenish and purple light extending round the horizon, even beyond the north [A glow heralds the beginning of the display]. Over the northeast quarter, the air to a height of 30 degrees had a dark opacity, which had the effect of arresting the light coming from beyond. [This is probably just clouds.]
>
> At 7h 30m PM an irregular obscure space began to form along the northern horizon. At 7h 50m a faint arch of white light made its appearance, resting on the horizon a little north of the east and west points, and culminating some distance below the pole star. This continued to rise until 8 PM, when its apex was within a few degrees of the pole. [Obviously this is what is now called an arc.]
>
> At 9h 20m a low luminous segment showed itself on the horizon beneath the arch. The latter now resolved itself into an array of bright streamers, with equidistant shadowy spaces between them. [A patch or surface aurora appeared and the arc began to develop into rays.]
>
> At 9h 30m the streamers had extended and grown brighter, while the low luminous segment, diffusing itself upward, had merged into the outer arch, which now reached nearly to the pole star. At this moment, the arch began to send off successive waves of light, rapidly following each other towards and beyond the zenith. In a few moments this wave movement gave way to more rapid and seemingly broken pulsations, flitting upwards in close succession through the northern, eastern and western quarters of the sky, and visible, though less distinctly, in the south. This wonderful appearance exhibited everywhere a convergency of the lines of motion towards a point considerably south of the zenith. [A corona had formed.]
>
> When these luminous phenomena were at their height, every spot to which the eye was directed, except the southern quarter near the horizon, was traversed by quickly successive flashes of white, greenish, and pale roseate light, all seemingly moving upwards.
>
> At 10h 30m the pulsating movement again extended over all the

> northern and part of the southern half of the sky. Innumerable waves of white, yellowish and purplish light chased each other from every quarter towards the magnetic pole, while the crimson flush spread wider and higher from the west.
>
> The various phases of the aurora recurred according to a somewhat uniform order of succession. First, the dark segment on the northern horizon took a regular arched form, and as it rose, became bounded above by a broad luminous curve, at the same time developing one or more bright concentric arches within. The streamers now shot forth from all parts of the luminous zone; and as these increased the upper arch faded away, as if it had expended itself in producing them. And now the lower arch took its place, to be obliterated in turn by a like seeming process of exhaustion. At length, one of the grander effusions of light coming on, the whole arch was broken up, and the dark segment below was reduced to a shapeless mass. Then there occurred a comparative pause in the phenomena, until the dark segment again took form, with its one or more luminous bands, and a like cycle of development was repeated.

In Bermuda (32 degrees 34 minutes) the intensity of the light stirred sleepers from their beds, and in Savannah, Georgia (32 degrees 5 minutes), the population was treated to a frenetic display in which intense pinks, golds, and purples rose upward from the east and west to an elevation of about 45 degrees before "dissolving." By 2 A.M., the aurora had succeeded in joining together to form a complete arc across the sky and then, an hour later, it gathered itself into a corona that "sent out bright fiery flashes in every direction."

The northern auroras marched downward, becoming visible in Key West, Florida (24 degrees 32 minutes), and Havana, Cuba (23 degrees 9 minutes); continued into the skies of the fecund realms of Cancer at 23 degrees north, before crossing Inagua, in the Bahamas (21 degrees 18 minutes).

One inhabitant, identified only as "a Spanish Mechanic," of Saint Jago de Cuba (20 degrees) passed the opinion that the locals would think the end of the world was at hand. A large proportion of people in Kingston, Jamaica (17 degrees 58 minutes), certainly thought so, believing the red sky signaled that Cuba was being consumed by fire. Although others postulated that these lights were the auroras, the majority dismissed the idea because the northern lights had never before been seen from the island. Sky watchers in Guadeloupe (16 degrees 12 minutes), West Indies, were treated to a pair of whitish arcs that passed a little to the west of the low-lying North Star, Polaris.

The aurora also visited Australia and the Pacific, although, because of their different time zones, it was the evening of 2 September when the heavenly lights arrived. Shortly after sunset, a rich pink light glowed in the southern skies over Kapunda, South Australia. The skies remained like this until around 9 P.M., when "a huge pillar of fire appeared in the west." As the Moon set on this scene, the aurora increased in intensity, more than compensating for the lost moonlight. A false dawn appeared, with the sky changing from more sanguine hues to bluish green. From this, the now familiar streamers of color shot toward the zenith.

A striking similarity in the reports is that most places attest that the auroras were only chased from the sky by the daybreak, although several places had the spectacle brought to a premature end by the appearance of rain clouds.

As for the cause of these glorious displays, the Victorians had no idea. Their only clue dated from 1741 when O. P. Hiorter, a Swedish graduate student under the tutelage of Professor Anders Celsius (after whom the centigrade temperature scale is also known), noticed a pronounced disturbance in compass needles whenever an aurora was overhead. Recounting the tale six years later, Hiorter wrote that when he told Celsius of the discovery, the professor said he too had seen such a disturbance but had decided not to mention it, in order to see if Hiorter would notice it.[2]

So the aurora was somehow magnetic but, beyond that, people knew very little. Scientists had been at a virtual standstill about auroras for over a hundred years. That all changed the day Carrington walked into Kew.

Although known as the Kew Observatory, the white stone building was actually situated in Old Deer Park, Richmond. George III had built the three-story mansion in 1768–1769 as his private observatory. In be-

[2] Commenting on this paper, A. J. Meadows and J. E. Kennedy noted wryly that "the relationship between senior research personnel and their assistants has not changed greatly down the years" (*Vistas in Astronomy* 25 [1982]: 420).

tween his infamous bouts of madness, he had—sometimes personally—set the time for the Houses of Parliament, Horse Guards, St. James, and elsewhere with observations made atop the balustraded roof, using a series of stone pillars across the park to tell when the Sun was precisely due south. The monarch had also personally ended John Harrison's long tussle with the Board of Longitude by supervising the testing of the H5 timepiece at Kew in 1772.

When Richard Carrington made his way up the long drive to the now-extended observatory, eighty-seven years later, there was good news and bad news waiting for him inside. The bad news was that no one at Kew had seen the solar flare or had even photographed the Sun that day. The latest photograph they had was from 31 August. However, and this was the good news, something strange had taken place on 1 September. Something that had made the Earth's natural cloak of magnetism quake.

The magnetic instruments at Kew had captured the disturbance. Each consisted of a compass needle hung on a silken thread in a darkened room. The needle pointed north, corraled by Earth's magnetic field. Any change in the field resulted in the needle moving. To record such movements, a ray of light shone onto the reflective needle, from where it bounced onto a slowly rotating drum. Around the drum was fixed a curve of photographic paper. Every day at 10 A.M., the Kew technicians changed this paper, which then snailed round at just three-quarters of an inch per hour so that twenty-four hours later, it had produced a tracing eighteen inches long. Any disturbance of the Earth's magnetic field caused the needle to quiver and a jagged line to be etched onto the paper.

Carrington was shown the scroll for 1–2 September. As near as could be deduced from the somewhat restricted scale of the tracing, the Earth's magnetic field had recoiled as if struck by a magnetic fist at exactly the same time as he had seen the flare. The abrupt part of the disturbance had lasted just three minutes but it had then taken the next seven to die back down to normal. This considerably raised the stakes. Providing they were not being fooled by mere coincidence, it seemed that Carrington's flare had somehow reached out across 93 million miles of void and struck the Earth. It must have been a dizzying and slightly terrifying moment. For two hundred years, astronomers had used Newton's laws of gravity to understand the interaction between the Sun and the planets. Astronomers had come to believe that gravity's gentle grip moved the Universe with predictable regularity, casting all humanity in the role of spectator to the grandeur of the clockwork cosmos. Now, however, the picture seemed as if it might be dramatically different. Earth was in the thick of things. Gravity was undoubtedly still the major player but sudden, unexpected magnetic strikes could occasionally steal the show. They were not just sharp jabs, either. Eighteen hours after the initial disturbance, Carrington was shown that the Kew needles again started moving, surpassing the strength of the 11:20 jab. This time, instead of a single punch, Earth began to suffer a sustained assault unequaled in the decades that Kew had been collecting data. In fact, on the day Carrington stood in Kew, the needles were still uneasy. The magnetic storm, although diminished, had by no means subsided.

As darkness fell on the evening of 2 September, the auroras were still raging, affording Europe its first view of the unprecedented light show. Christiania, Norway; Durham, Preston, Nottingham, Grantham, London, Clifton, Aldershot, Brighton, all in England; Paris, France; Brussels, Belgium; Prague, Rzeszow, Vienna, Mitterdorf, all in the Austrian Empire; all over Switzerland; Rome, Italy; all saw auroras. Other places, such as Sweden and across Russia, recorded violent thunderstorms. Although northern Asia was similarly overcast, magnetic instruments there also registered a mighty disturbance.

One of the lowest latitude auroral observations came from this night also. It was reported in the *Gaceta del Estado*, from La Union, San Salvador, just 13 degrees 18 minutes north of the equator:

> About 10 o'clock, a red light illuminated all the space from north to west, to an elevation of about 30 degrees above the horizon. The light was equal to that of daybreak, but was not sufficient to eclipse the light of the stars. The sea reflected the color, and appeared as if of blood. This lasted until three in the morning, when a dense black cloud arose in the east, and commenced to spread over the colored portion of the heavens, presenting a most curious spectacle; for in the parts where the cloud was not dense enough, the red light shone through, and formed a thousand fantastic figures, as if painted with fire on a black ground.

The sanguine imagery was repeated in a report from the nearby city of Salvador, where "the red light was so vivid that the roofs of the houses and the leaves of the trees appeared as if covered with blood."

As these and many other reports slowly filtered around the globe, it became obvious that something extraordinary had happened to Earth. The planet had been an unsuspecting participant in a celestial event of enormous proportions. And Richard Carrington may have witnessed the birth of the mayhem. But could anyone actually back up his story of the tumultuous outburst? No matter how eloquently, diagrammatically, or mathematically he described what he had seen, until he could find someone willing to vouch for him there would undoubtedly be skeptics. That was the nature of science—no acceptance without proof, and the more extraordinary the claim, the more extraordinary the proof required. Carrington's calculations suggested that the extent of the event was not just extraordinary; frankly, it smacked of the unbelievable. Carrington did have one asset: his reputation. He was known for his meticulous attention to detail, which bordered on the obsessive. Even the Astronomer Royal, George Airy, consulted him on the precision of observations taken from the hallowed domes of Greenwich.

After Carrington's Kew visit, the astronomical grapevine circulated his claims. The Reverend William Howlett of Hurst Green, Kent, had observed the Sun that day, but began at noon, more than half an hour after the dramatic event. Nevertheless, a bona fide astronomer turned up who had been observing the Sun on 1 September 1859. He was R. Hodgson, Esq., of Highgate, himself a respected solar observer who had invented a special eyepiece with which to safely observe the Sun's fearsome light, and a Fellow of the Royal Astronomical Society. No doubt to Carrington's great relief, Hodgson also considered that he had witnessed "a very remarkable phenomenon." Upon learning this, Carrington insisted that they exchange no further information. Instead, they would both present their considered accounts at the next appropriate meeting of the Royal Astronomical Society.

In the intervening time, reports arrived that made it apparent that the auroras had possessed a sinister side, too. The beguiling lights had somehow disabled the telegraph system, wiping out communications across the world, just as surely as if someone were to pull the plug on today's Internet. Like the modern reliance on the electrical ether of the

World Wide Web, business at that time used the telegraph for trading stocks and shares, governments relied on it for intelligence and news, and individuals used it to keep in touch with loved ones. Yet for days after Carrington's flare, nature refused to allow these arteries of information to flow freely.

At its mildest, the disruption was an inconvenience, as when the aurora made the incoming message bells ring spontaneously in Paris and other places. At its worst, the aurora presented a danger to life and limb. In Philadelphia, a telegrapher was stunned by a severe shock while testing his communications equipment. Those stations using the Bain or chemical system, which used the electricity on the line to mark sheets of paper and thus record the incoming messages, were put in the gravest danger. When the currents surged powerfully enough, the paper would catch fire, engulfing the stations in choking smoke. In Bergen, Norway, the appearance of the aurora conjured such strong electrical currents that operators had to scramble to disconnect the apparatus, risking electrocution to save the equipment from destruction.

Science had to solve the mystery of what caused the aurora.

On 11 November 1859, the Fellows of the Royal Astronomical Society gathered in anticipation at Somerset House, London. The audience of gentlemen, no doubt dressed in the fashionably long frockcoats and adorned with the increasingly intricate neckties of the day, listened with rapt attention as first Carrington and then Hodgson offered their accounts. While giving his testimony about the Promethean event, Carrington showed an enlarged copy of the precision drawing he had made on the day. Afterwards he lodged the artwork at the society's rooms, so that members not present at the meeting could inspect it at their leisure. He then took his place in the audience and listened as Hodgson told his story, doubtlessly anxious to learn whether their accounts would tally.

In broad agreement with Carrington, Hodgson told how he had been surprised by the appearance of a bright star of light, much brighter than the Sun's surface. He described how the dazzling light illuminated the edges of the adjacent sunspot rather like the proverbial silver lining seen around clouds. His timings also matched Carrington's. However,

Hodgson confessed that his surprise—and the fleeting nature of the outburst—had prevented him from making an accurately measured drawing. Instead, he had executed a sketch. It too was left behind for private inspection after the meeting, and it was noted in the editorial comments of the society's journal that it was a well-executed drawing that excited much interest at the meeting. For unknown reasons, though, the journal did not reproduce it alongside Carrington's and it now seems lost.

At the end of the discourses, no Fellow could be in serious doubt that something unprecedented had indeed taken place on the Sun or, more likely, just above it, for Carrington had convincingly argued that since the Sun's surface had displayed no difference before and after the event, the flare must have taken place high above the sunspot group. As for the putative link between the flare and the auroras, there was considerable debate. Both men had mentioned these two features; Carrington even showed photographs of the Kew charts, pointing out the magnetic jolt at the time of his flare and then drawing attention to the subsequent and incredibly powerful magnetic storm that coincided with the auroras. He must therefore have thought it important, but on the day he remained a paragon of scientific skepticism, cautioning his audience that while the contemporary occurrence may deserve further consideration, "One swallow does not make a summer."

One reason for his overt caution was that neither Carrington nor anybody else could conceive of a mechanism by which the Sun's explosive force could be conveyed to Earth. If the link were real, it would require new physics. Without this, all Carrington really had were two pillars on either side of an immense ravine but nothing with which to bridge them. Quite clearly, he was not about to commit the classic amateur blunder of drawing a mammoth conclusion from a single, solitary example.

With the luxury of a century and a half's hindsight, we can now see that the Carrington flare was a tipping point in astronomy. The sudden demonstration of the Sun's ability to disrupt life on Earth catapulted astronomers into a headlong race to understand the nature of the Sun. Previously, such investigations had been a backwater of astronomy with the main science concentrating on charting the stars to help navigation. In the same year that Carrington saw the flare, a breakthrough in the analysis of light was made in Germany. It gave astronomers the means with which to investigate the Sun's composition. Once they had used these techniques on the bright Sun, so they adapted them to investigate the stars, and traditional astronomy began its transformation into present-day astrophysics.

At the fulcrum of this change is the realization that the Sun's magnetic energy could strike the Earth, proving that the celestial objects were connected in ways previously unimagined. Yet Carrington's flare and the subsequent magnetic mayhem were not the first events to force astronomers to consider a link between the Sun and the Earth beyond mere sunlight. Over half a century earlier, master astronomer William Herschel had presented a series of ideas to the Royal Society about the nature of the Sun. During the final one, he explained how he had seen something in the fluctuations of wheat prices that made him think of sunspots. The mere suggestion had caused uproar and Herschel was ridiculed for his thoughts.

TWO

Herschel's Grand Absurdity, 1795–1822

In the late 1700s, more than half a century before Carrington's flare, public opinion about the Sun's nature, even its purpose, was split along an irreconcilable divide. On one side stood "fanciful poets" who sought to make the Sun "the abode for blessed spirits." On the other were "angry moralists" who pointed out that the Sun "is a fit place for the punishment of the wicked." On 18 December 1795, the fifty-seven-year-old William Herschel rounded on both viewpoints, pronouncing them the product of opinion and vague surmises. He claimed that his own observations gave him the authority to propose a third viewpoint based solely upon "astronomical principles": the Sun was not a place for spirits, either of the deserving or the wretched; instead it was a place of living creatures, a vibrant world every bit as inhabited as the Earth.

Many were already suspicious of Herschel's brand of astronomy, believing that he hovered in the slim space between madman and genius. He himself once quipped that the astronomers would have him carted off to Bedlam if for nothing else than the magnifications he regularly used for observing the stars. This latest claim certainly did nothing to increase his standing with the detractors.

Part of Herschel's credibility problem lay in the fact that his homemade telescopes were generally superior to anything the professional astronomers had at their disposal. Having become used to the amorphous shapes that resulted from their own imperfect telescopes, they nurtured a particular grievance against Herschel's claim to see stars as round objects at high magnifications. Herschel once found himself seated at dinner next to physicist Henry Cavendish, a taciturn individual who would quietly brood upon contentious issues and communicate with his female servants only by written note because he was too shy to actually speak to them.

Characteristically, he said nothing during the beginning of dinner. Then he turned. "I am told that you see the stars round, Dr. Herschel."

"Round as a button," came the reply.

Cavendish returned to silence and the dinner continued. Toward the end of the meal, he turned again. "Round as a button?"

"Exactly round as a button," said Herschel. And so the conversation between the two men ended.

Even the telescopes at the Royal Observatory in Greenwich were poor in comparison to Herschel's and the Astronomer Royal, Dr. Neville Maskelyne, disbelieved Herschel's claims of magnifying stars by thousands of times—especially since he himself was making do with magnifications of just one-tenth that power. Following a dinner with Maskelyne, a friend wrote to Herschel: "You have arrived at such perfection in your Instrument, or at least have dared to apply it to such uses and so have overleaped the timid bounds which restrain modern astronomers, that they stand aghast and are more inclined to disbelieve than to admit such unusual excellence."

Herschel replied plaintively, "My observations will stand much in need of the protection of some kind gentleman well known in the astronomical line . . . give me leave to beg of you (for the love of Astronomy's sake) to lend your assistance, that such facts as I have pointed out may not be discredited merely because they are uncommon."

For most of his life Herschel successfully fought off the criticisms, insulated by a unique achievement that placed him without peer; he was the only person in all of history to have discovered a planet. The seventh planet of the solar system, Uranus as it became known, appeared in his homemade telescope in 1781, while Herschel was observing from the back garden of his home in the English spa town of Bath, accompanied by his faithful scribe and sister, Caroline. The discovery made the forty-three-year-old Herschel a celebrity and brought him royal patronage from George III. It also forced his maverick brand of science onto the professional astronomers of the day.

Instead of concerning himself with measuring the position of stars to construct more reliable navigation tables, Herschel concentrated on discovery. He loved uncovering the form and quantity of different celestial "species," as he called them. His approach owed a debt to the scholars of natural history, many of whom he had rubbed shoulders with at the Bath Philosophical Society. As the natural historians set about studying the flora and fauna of the world they lived in, building up the taxonomy of classes and species, so Herschel decided to do the same with the Universe.

This break with traditional observing goals, almost certainly exacerbated by his status as an amateur astronomer, accumulated detractors. They spread pervasive rumors that he had been no more than lucky in his discovery of Uranus. Herschel's supporters, mainly drawn from the ranks of natural historians rather than astronomers, dismissed the malcontents as "jealous barking puppies." Herschel himself defended his find with vigor, insisting that his unprecedented discovery was a natural consequence of his heavenly survey. On one occasion he wrote that he had "gradually perused the great volume of the Author of Nature and was now come to the page which contained a seventh planet."

Herschel's astronomical agenda had a grand goal: to discern the complex interplay of the heavenly objects by incessant observation. It was a sentiment that resonated with the burgeoning romantic literary movement of the time, who believed that to see and experience a thing was a major part of coming to know that thing. They also argued that mankind could not be removed from the equation, as natural philosophers often strived to do in their quest for objectivity. Herschel melded both qualities, often beginning from wholly objective observations before spinning them into a richly detailed picture that he hoped others would discuss.

On 18 December 1795, Herschel did something that no one before him had attempted. He began a series of lectures that he hoped would spark a grand discourse on the nature of the Sun and the precise links it shared with the Earth. At this time in history, there were no individual societies for the discussion of the separate sciences. It would be another twenty-five years before the Royal Astronomical Society was inaugurated, so Herschel's platform was the Royal Society, the learned organization for all natural philosophers of standing, regardless of their individual areas of expertise, one into which Herschel had been ushered following his discovery of Uranus.

The society officially began in 1662 when a meeting group of scholars obtained a charter for their activities from Charles II. Now, 130 years on, the Royal Society met in the newly completed Somerset House, an architectural marvel that backed onto the Thames and was fronted by London's Strand. Surrounded by the oil paintings of previous Fellows, the existing ones sat on wooden pews beneath chandeliers and an ornate paneled ceiling, listening as Herschel added his unique brand of astronomical analysis to more than 180 years of solar observations.

Telescopic observations of the Sun had begun around 1610, when Galileo had noticed the Sun setting over the Italian city of Padua. Hanging low in the sky, the burning intensity of the Sun's glare had been substantially reduced by thin clouds or mist. Galileo took his telescope and studied the luminous orb. He noticed dark spots upon its surface and used them to measure the rotation rate of the Sun, about twenty-five days according to Galileo's estimate.[1]

Others with telescopes were also noticing sunspots: Johann Goldsmid in Holland, Christoph Scheiner in Germany, and Thomas Harriot in England, all pondered the transitory nature of the markings. Scheiner believed that the spots were the silhouettes of previously undiscovered planets. Galileo railed against this idea, using a series of observations to show that sunspots display a peculiar speed-up, slow-down motion as they cross the Sun. When spots appear on the edge of the Sun, Galileo observed them to speed up gradually until they reached the solar center; thereafter they slowed down again and finally crawled out of view, over the opposite edge. Galileo correctly reasoned that this is exactly the behavior of something affixed to the surface of a spinning ball, whereas a planet would travel in front of the Sun at a constant speed. If this argument were not convincing, Galileo also called on observations that showed sunspots growing or shrinking in size. Some had even

[1] It is a myth that Galileo went blind as a result of his solar observations. Galileo's eyesight faded when he was seventy-two owing to cataracts and glaucoma. His solar observing took place twenty-five years earlier. Although he originally viewed the Sun directly at sunrise or sunset, Galileo soon switched to projecting the image of the Sun onto a flat board.

dwindled to nothing under his gaze. How could an intervening planet behave like that?

Certainly by the time Herschel delivered his lecture, no one of a learned disposition believed that sunspots were anything other than features on the Sun. There were a number of ideas about exactly what the sunspots might be. Galileo had proposed that they were dark clouds, drifting in the solar atmosphere. Other early astronomers believed that the Sun was a gigantic natural furnace, so the dark spots must be slag floating on top of its incandescent fluids. Bizarrely, the appearance of three bright comets in 1618 bolstered this view. By coincidence, the Sun was unusually free of large spots that year, and some argued that the cinders and burned effluent that would usually create sunspots had instead been blasted into space, where they had become the comets.

Opinions changed in the early 1700s, largely because Isaac Newton had written in his 1704 book, *Opticks* "And are not the Sun and fix'd Stars great Earths vehemently hot." This led thinkers to the idea that buried somewhere inside the Sun's incandescent exterior was a planetary body, perhaps like Earth. Following this line of reasoning, volcanic eruptions became a favorite explanation for the spots. They were envisaged to be the belched smoke that presaged a violent eruption. Others modified this view to suggest that they were not volcanoes but merely mountains revealed by the occasional ebb and flow of the Sun's fiery oceans.

Following the briefest summary of these various ideas, Herschel set about an exhaustive description of his own solar observations. Most previous observations had been taken using small telescopes to reduce the sunlight collected to tolerable levels. Astronomers might then hold smoked glass in front of their eyes to further reduce the intensity. The trouble was that small telescopes could not see details as easily as large telescopes. However, if you tried to use a larger telescope and compensate with smokier glass, the concentrated sunlight blistered the smoky surface until it bubbled away in patches, allowing piercing rays to shoot through and injure the eye. Or worse still, the glass would shatter under the heat, often with the astronomer's eye mere inches away.

Herschel's breakthrough was to hone a telescope mirror but not to polish it. In those days mirrors were not made of glass with a thin layer of silver or aluminum. Instead they were constructed from speculum metal, an alloy of copper and tin to which was added a dash of arsenic to help the metal take a polish. Herschel's innovation provided a poorly reflecting surface that naturally cut down the ferocity of the Sun's rays, allowing him to use larger mirrors and thus see more detail than anyone before. Instead of an indistinct glowing mass, the surface resolved itself into a mottled expanse that he likened to orange peel. He also saw that large sunspots were depressions that sank beneath the luminous layers, convincing him that they were "openings" through which the actual surface could be glimpsed.

However groundbreaking his ideas seemed, Herschel was not the first to advance these views. Dr. Alexander Wilson, the former professor of astronomy at the University of Glasgow and the late father of Herschel's friend Patrick Wilson, had reached the same conclusion in 1769, twenty-six years earlier. All Herschel had really done was corroborate Wilson's view. Yet, Herschel had chosen not to mention Wilson's priority in these claims. This omission would not go unchallenged.

Oblivious to the impending controversy, Herschel proposed that the luminous surface was not an ocean but an atmosphere composed of two distinct vapors: one transparent, the other bright, or lucid as he termed it. In his view, the sunspots were fostered by the landscape of the solid world beneath, creating eddies that opened temporary gaps in the brightness.

Although he had no explanation for what the lucid matter might be, he pointed out that there were precedents. On Earth, the atmosphere occasionally shone with the auroras. Observers of Venus sometimes reported seeing an ashen light mottling its dark side.[2] Lacking the necessary knowledge of matter's atomic structure to explain such fluorescence, most natural philosophers simply thought that luminous matter must contain some as yet undiscovered chemical ingredient that glowed under the right conditions. Whereas on Earth and Venus atmospheric luminosity was a temporary phenomenon, Herschel argued that on the Sun it was clearly the most natural state of the solar atmosphere.

[2] Since its discovery in 1643, the ashen light of Venus has proved highly controversial. The 1990 fly-by of Venus by NASA's *Galileo* spacecraft provided what some believe to be an explanation. The surface of Venus is so hot (over 400 degrees Celsius) that the rocks glow. Sometimes the all-encompassing Venusian clouds thin out and allow the glow to escape into space, for eagle-eyed observers on Earth to see. Others believe it is an atmospheric effect called "airglow," which is similar to earthly auroras. Still others maintain that the ashen light is a myth, seen only by those with good imagination or poor optics.

Herschel pressed on to his ultimate conclusion: that the Sun was richly stocked with inhabitants. He argued that it was actually an inevitable consequence of his having "proven" that the Sun was a type of planet. Just as explorers on Earth found human societies on even the remotest shores of the New World, so astronomers believed that all planets were similarly inhabited. If the Sun were a planet, then it had to be a living planet. He then asked his audience a rhetorical question. Since the Earth, at a distance of 93 million miles from the Sun, was warmed across its surface, would not the Sun's surface be scorched beyond all conception?

Herschel thought not. In fact, he wasn't convinced that the Sun was hot at all. As evidence he offered the experiences of mountaineers and balloonists who uniformly reported a drop in temperature as they ascended to great heights, which Herschel said took them closer to the Sun. Totally misunderstanding the role of altitude in atmospheric temperature, Herschel believed this implied that the Sun was fundamentally a cool body. He argued that heat was not an innate property of sunlight, but the result of its interaction with a susceptible piece of matter. If the Sun's solid surface largely lacked this quality, it could be a perfectly temperate place.

As a final argument, he appealed to common sense (!) and cautioned the audience not to fall into the same trap as the supposed inhabitants of the Moon, who might be looking down at the Earth and assuming it was only there to hold the Moon in orbit and reflect more warming sunlight in its direction, missing the point altogether that Earth was an abode of life all its own. Thinking that the Sun's sole purpose was to corral the planets with gravity and supply them richly with warmth would be a similar mistake, according to Herschel.

Soon after the lecture, a letter arrived at Herschel's home in Slough, where he had moved to attend the King's observing whims at the nearby castle of Windsor. The letter was from Patrick Wilson, who had succeeded his father to the astronomy professorship at Glasgow. Wilson asked Herschel to explain why he had claimed the discovery of sunspots as openings when Wilson senior had published this same conclusion decades earlier.

Herschel destroyed the letter, as was his custom with communications that distressed him or contained comments that he judged might become embarrassing for the writer. His reply survives because Caroline transcribed it into the family's letters book. Herschel implied that he had not known of Wilson's paper on the subject but that now he had read it, he "avowedly disclaimed every merit as a first discoverer." He then explained that he had neglected to survey the recent literature on sunspots because he wanted to avoid a fight with the German astronomer Johann Hieronymus Schröter, who, he confided, had recently written "a tedious treatise on the solar spots." In talking about recent ideas, Herschel would have felt duty bound to take issue with those ideas, thus inviting an argument because Schröter showed "a disposition to take hold of every opportunity to defend his erroneous as well as his good communications."

This was perhaps the first sign that the feisty nature Herschel had displayed defending Uranus was beginning to ebb away. At age fifty-seven, he was beginning to fill out and slow down. His observing logs show that he no longer spent every possible night with his sister sweeping the starry vault for new discoveries, as had once been their combined passion. He had made a late marriage and now had a three-year-old son. He was finding it difficult to make ends meet and moved to cheaper accommodation after his landlady insisted on rent increases every time he erected a new telescope. His royal pension was a niggardly £200 a year. He supplemented this by building telescopes for the rich and noble, but maintaining the craftsmanship for which he was renowned was time consuming. In his letter to Patrick Wilson, he lamented the fact that he was engaged in the construction of a thirty-foot-long telescope for the king of Spain and that meant there would be "no possibility of leaving the workmen this twelvemonth." On top of everything, visits to his house and observatory were now a social fixture for the gentry passing through Slough and Windsor. Playing host, no matter now much he enjoyed the attention, sapped his time and energy.

When he did get out at night, he concerned himself with the brightness of stars. The stars that particularly captured his attention were the handful that changed their brightness. Knowing of only one variable quantity on the Sun, the number and size of sunspots, Herschel proposed that variations in stellar brightness must be due to the appearance of large spots in great number. He wondered what would happen if the Sun suffered a similar outbreak. Surely the resultant drop in sunlight would be devastating to life on Earth? In his search for an answer Herschel looked to the work of the natural historians and claimed that many phenomena in natural history seemed to point out past changes in our climate.

Although he failed to commit his precise reasoning to paper, he was almost certainly thinking about the growing belief that the landscapes of the Earth had come gradually into being, maturing into their present forms over the ages rather than appearing in final form, as a literal translation of the Bible demanded. As the natural philosophers investigated the rocky layers exposed at cliff faces and in mineshafts, they found that things in the past had been different from today. Fossils were a particularly good illustration. Buried in the storm-wracked limestone cliffs of southern England were the remains of creatures that resembled tropical crocodiles. Could England really have been a sultry realm more like the equatorial regions?

We now know that rocks are shunted around the Earth over eons by the forces at work inside our world, but for Herschel such ideas, not to mention the technology needed to measure the drift of continents, lay centuries in the future. The only interpretation open to him was that England had once been much warmer. He wrote that "perhaps the easiest way of accounting for [apparent changes in climate] may be to surmise that our Sun has been formerly sometimes more and sometimes less bright than it is at present. At all events, it will be highly presumptuous to lay any great stress upon the stability of the present order of things."

As a result of his thoughts about climate change, Herschel's investigation of the Sun gained renewed momentum. He set about adapting a telescope with a mirror fully nine inches in diameter and hit a snag almost immediately. The focused heat cracked every one of the darkened slips of glass that he used to reduce the glare. Always the technological innovator, Herschel began experimenting with colored glass to perfect the ideal solar eyepiece. He discovered that while red glass stopped most of the light from the Sun, it left an intolerable sensation of heat on the eye. When he held a thermometer in the place of his eye, the instrument jumped 29 degrees immediately. He snatched it away lest the rapidly expanding mercury ruptured the glass tube. Green glass, on the other hand, cut out the heat but let through too much light.

At the time it was thought that different colors created equal quantities of heat. Herschel's colored glasses showed this was not the case. Intrigued, he began an investigation. He placed a prism in a sunlit window and used it to cast a rainbow spectrum of colors onto a pivoted board in which he cut a long slit. Underneath the board he set three thermometers, one of his own and two borrowed from the mollified Wilson. By maneuvering the board he could vary the color that fell through the slit onto the mercury bulbs below. He worked through the colors, from blue to yellow to orange, recording just how much each raised the temperature of the thermometers. By the time he arrived at the red light, the temperature was still rising. There could be no mistaking that red light really did possess more heating powers than the other colors.

He gave his first announcement to the Royal Society on 27 March 1800, sending more papers to be read four weeks later. He presaged his surprising conclusions with some defensive comments, reminding those assembled that it was sometimes the duty of natural philosophers to doubt the things commonly taken for granted. With his audience forewarned, he asserted that heating ability was unevenly distributed among the colors of light, with the blue rays almost devoid of the power.

As he had feared, not everyone believed him. His results were treated in some quarters as little more than the ramblings of a fool. A petulant paper was published by a Mr. Leslie in the *Journal of Natural Philosophy, Chemistry and the Arts*, claiming that Herschel had made an amateurish mistake and recorded only the rising ambient temperature of the room. Herschel rose above the attack, his equanimity no doubt boosted by the unqualified support of Sir Joseph Banks, president of the Royal Society. Banks wrote a deferential letter to Herschel, requesting a face-to-face meeting in which to discuss the new work. Perhaps mindful of the criticism Herschel was receiving, he reassured the astronomer that although "the public will appreciate as they see fitting the value of your discovery; for my own part I hope you will not be affronted when I tell you that, highly as I prized the discovery of a new Planet, I consider the separation of heat from light as a discovery pregnant with more important additions to science."

Clearly inspired, Herschel pressed on with his work, using the Royal Society again as his vehicle to describe an even more audacious conjecture: not only were some colors unconnected with heating, but most of the Sun's true heating power was carried on invisible rays, residing beyond the red end of the visible spectrum. He described again placing a prism in the window. This time he darkened the window around the prism with a heavy green curtain so that there could be no stray light to affect the thermometers. The colored spectrum shafted from the prism onto a tabletop covered in white paper. He then moved the thermometers from one color to the next, taking readings as he went.

To stifle his detractors, he placed one thermometer in line with the other two, but to the side of the colors, so that it remained unlit. With this arrangement he could monitor the ambient temperature of the room, checking that the colored thermometer readings really were higher. This was his master stroke, and it allowed him to be confident about his next move. Having taken readings for all the colors, he slid the thermometers above the discernible edge of the red light. Instead of returning to the ambient temperature of the room, the thermometers rose again. In fact, they rose far beyond any reading that occurred when they were in the colors. The offset thermometer, on the other hand, continued a steady reading. The conclusion was inescapable: the Sun's heat was mostly carried on invisible rays that behaved like light but could not be discerned by the eye.

Herschel wanted to refer to them as caloric rays in his announcement, but Banks persuaded him against this because the system of chemistry, from which Herschel had borrowed the phrase, was under attack by the French. Banks believed that to derive a name from a beleaguered theory might predispose more people against the discovery itself. So Herschel adopted Bank's suggestion of radiant heat as an interim. It then took eighty years before the scientific community settled on the term "infrared" for Herschel's invisible heat rays.

The discovery of radiant heat destroyed Herschel's ideas that the Sun might be a cool body but did set him thinking again about solar variability and the disasters this might prompt on Earth. There was sound social context for this worry. The French revolutionary wars were raging across Europe, transforming into the Napoleonic Wars as they went. Britain had been at war with France since 1793 and was isolated, unable to avail itself of imported grain from Europe. This was good news for British landowners, who now monopolized the English market, but terrible for the ordinary folk. All that stood between them and starvation was bread. If the price of grain rose, there was a real chance their meager wages would not stretch to buy sufficient food, especially if they had a dependent family.

Herschel saw a direct analogy. The Egyptians knew that the Nile sometimes flooded and that the fortunes of their harvests were tied to those floods. Yet the floods were irregular, influenced by natural forces beyond Egyptian control. Recognizing this, they sought to divine the signs of an impending bad flood season, so that provision could be made. On 16 April 1801, he implored the Royal Society: "Should not an understanding of the Sun's glorious output be similarly studied, and provision made for those years when sunlight and heat are deficient and the harvest bad?" To Herschel, this was no longer a matter of mere scientific curiosity. The Sun's light and heat were the ultimate font of all earthly existence. It was imperative to understand what level of permanence could be ascribed to the rays. He set about a definitive study of Earth's true link with the Sun.

Throughout the previous decade, he had been amassing reams of notes about the sunspots he had seen. As he perused the data, something struck him. Between 5 July 1795 and 12 February 1800, there had been many days on which spots had been absent from the disk. Then, they had suddenly returned with their usual profligacy. So the spots were not consistent. He wrote, "It appears to me, if I may be permitted the metaphor, that our Sun has for some time past been laboring under an indisposition, from which it is now in a fair way of recovering."

It seemed to Herschel that his prediction of the Sun's variability was coming true before his eyes. But was this an isolated blip or a repeating cycle? He needed more data. Scouring past journals in search of earlier sunspot observations, he could find only scant periods of real interest by his astronomical forebears. He publicly rued this dearth of consistent observations, for it seemed only to confirm the taken-for-granted attitude toward the Sun that he now fought against.

Despite the paucity of data, he persevered in his efforts and eventually identified five previous periods in which he believed the sunspot numbers had declined: 1650–1670, 1676–1684, 1686–1688, 1695–1700, 1710–1713. But how could he check to see if there had been a change in the Earth's climate during these years? There were no systematic meteorological records to call upon; such endeavors were only just beginning. He needed to think of something that relied on the climate and was of such importance to society that records about it would be kept. Finally, he realized he needed look no further than the very thing that had precipitated his concern: the price of wheat. He turned to Adam Smith's classic work of 1776, *Wealth of Nations*, for the data. Herschel thought the concept perfect. In clement years, the grain harvest would be bountiful and the dictates of supply and demand would ensure a low price. In poor-weather years, the price would be correspondingly higher. All he needed to do was compare the wheat prices to the sunspot records. When he did so, he got a surprise.

If he thought that sunspots reduced the Sun's light and heat, thus depressing the weather, Herschel's tentative conclusion indicated the exact opposite. A dearth of sunspots seemed to correspond to higher wheat prices, and therefore presumably poorer weather. Periods of greater sunspot numbers somehow seemed to induce better seasons and more bountiful growth.

To explain this, Herschel appealed to his invisible rays of heat and suggested that clouds of a transparent "empyreal gas" were welling up through the luminous solar atmosphere, creating the openings seen as sunspots and releasing their quota of invisible heat rays into space, warming our planet and transforming our weather. In other words, the sunspots did not impede the outflow of heat but were the result of its copious emission. Having proposed the idea, he called upon others to take up the challenge of observing the Sun and test his theory. It was the boldest of all his discourses on the Sun, designed as a rallying cry for astronomers to put their science to a bold new use. No longer did he believe it was enough to chart celestial objects; he wanted astronomers to debate the very nature of them. However, his words fell on entirely deaf ears. Those who did listen decided not just to criticize him but also to ridicule.

The *Edinburgh Review* published a vicious assault on Herschel and his science. Henry Brougham, an educated Scottish reformer, derided Herschel's zeal for classification, calling it an "idle fondness." Believing that the astronomer did not make enough effort to fit his observations into the existing framework of natural philosophy and so rushed to coin new terms with nebulous definitions, Brougham wrote, "The invention of a name is but a poor achievement for him who has discovered worlds." The article's coup de grace contained the cruelest sentiment of all: "To the speculations of the Doctor on the nature of the sun, we have many similar objections [about his invented nomenclature] but they are all eclipsed by the grand absurdity which he has there committed; in his hasty and erroneous theory concerning the influence of the solar spots on the price of grain. Since the publication of Gulliver's voyage to Laputa, nothing so ridiculous has ever been offered to the world."

Brougham was referring to Jonathan Swift's *Gulliver's Travels.* Published in 1726, it was a satirical series of short stories designed to mock Britain's growing fascination with the customs and rituals of far-flung societies. The most famous story concerns Gulliver's encounter with the miniature people of Lilliput, but Brougham saw more parallels in the voyage to Laputa, in which Gulliver encountered a race of people who never enjoy a minute's peace of mind because their thoughts are continually full of apprehensions that the celestial bodies will change. One of Swift's examples reads, "That the face of the Sun will by degrees be covered in its own effluvia, and give no more light to the world."

In all probability, Brougham's lambasting was the backlash to a controversy that Herschel himself had started while simultaneously studying the Sun: a controversy that appeared to paint Herschel in the colors of hypocrisy. Ever since Herschel's discovery of Uranus, astronomers had been gripped by a feverish desire to find more planets. On 1 January 1801, the Italian astronomer Giuseppe Piazzi claimed to have joined Herschel's exalted rank by discovering another planet. Everyone believed the claim, except Herschel.

Giuseppe Piazzi had founded the observatory in Palermo, Sicily, in 1780. Being the most southerly of the European observatories, he had access to regions of the sky that no other European astronomer could see, and he spent his time compiling star catalog of these uncharted regions. While patiently engaged in this activity on the first day of the New Year, 1801, he recorded one faint star among many. He returned to the starscape the next night to check his measurements and found that it had moved. He tracked it over several nights, watching its passage through the firmament, and on 24 January wrote to a number of astronomers making his discovery known. Emulating Herschel's Uranus announcement, Piazzi claimed to have discovered a comet, but he confided his true aspirations to fellow countryman and astronomer Barnaba Oriani of Milan, writing, "I have announced this star as a comet, but since it is not accompanied by any nebulosity and, further, since its movement is so slow and rather uniform, it has occurred to me several times that it might be something better than a comet. But I have been careful not to advance this supposition to the public."

German astronomer Johann Elert Bode was in no doubt that Piazzi had found a planet and used his position as director of the Berlin Observatory and editor of the prestigious German astronomy journal *Berliner Astronomisches Jahrbuch* to trumpet the discovery across all Germany and beyond. He also claimed to have predicted the planet's existence. Back in 1768, at the eager age of nineteen, he had published a simple mathematical expression that, with remarkable ease, predicted each planet's distance from the Sun. He neglected to mention that fellow German astronomer, Johann Daniel Titius, had first formulated the expression in 1766. The Titius-Bode law predicted a planet between Mars and Jupiter, where astronomers knew of nothing but 300 million miles of void. Bode thought Piazzi's "comet" was the missing world, all that was needed to confirm this was a computation of its orbit.

Infuriatingly, the planet passed behind the Sun, masking it from view, before the necessary observations could be made. Astronomers had to wait until the end of the year for it to travel far enough from the Sun to again be visible in darkness. To their delight, its orbit was sitting neatly at the distance predicted by Bode. The new world was called Ceres.

If this were not wonder enough, on 28 March 1802, Dr. Heinrich Wilhelm Matthäus Olbers, a physician who moonlighted as an astronomer, discovered "another Ceres," following a similar orbit. For Bode it was an embarrassment of riches. If there were two planets between Mars and Jupiter, it threw out the numbering that was necessary to make his law work. He wrote to Herschel, stating that Ceres was the true fifth planet and that Pallas, as it had been named, was nothing but a celestial hanger-on—a remarkable comet snagged somehow by Ceres, or an exceptional planet beyond the orderly run of things. Olbers retaliated, writing to the Royal Society that "Pallas is a planet; the own sister to Ceres, not inferior to her in dignity and importance."

Herschel believed neither point of view. Using the extraordinary magnifications that his telescopes made possible, he had measured the diameters of these two supposed worlds and found them wanting. For Ceres, he computed just 161.6 miles, and for Pallas, about 147 miles. Johann Hieronymus Schröter, whom Herschel had avoided antagonizing several years before, had used other instruments and came up with a respectable 1,624 miles and 1,425 miles for the pair (about the size of the inner planet Mercury). Although both estimates are wrong by modern standards, Herschel's are the closer. Ceres is 583 miles across and Pallas 331.

With their diminutive size as his ammunition, Herschel set about destroying their claim to planetary status. He pointed out that they were virtually indistinguishable in appearance from stars and proposed the name asteroids, meaning starlike. Even though he knew full well that these objects could not be stars—he agreed with Newton that stars were similar to the Sun, just viewed from a much greater distance—he argued that the term asteroid was perfectly representative of these objects.

Letters of protest began arriving from Bode, Olbers, and Piazzi almost at once. On 17 June 1802, Olbers wrote again, this time with a compromise. What if these miniature planets were fragments of a once noble world that had been shattered by cosmic forces beyond rational credulity? There had been speculation among astronomers about the devastation a comet might cause if it were to strike a planet. Could they now be finding the evidence of such a collision? Herschel did not answer. On 4 July, Piazzi wrote suggesting the name "planetoids" for Ceres and Pallas. Quite reasonably, the Italian pointed out that asteroid was more suitable for little stars. Herschel ignored this suggestion, too. Shortly afterwards, Lord Brougham published his bruising appraisal of Herschel in the *Edinburgh Review* with its telling line, "The invention of a name is but a poor achievement for him who has discovered worlds."

In the face of such criticism, Herschel became more entrenched. A few months later he escalated the situation when a third miniature world was discovered. He investigated the new body, Juno, telescopically and commented that the term asteroid was perfectly suitable for it as well. While no one could deny the resemblance, to saddle these objects with a misleading name was an uncharitable act by someone whose place in history was secure. His fellow astronomers now began to place harsh criticisms of his work in print. In 1804, Bode published a translation of Herschel's sunspots–wheat-prices paper in the *Jahrbuch*. As was common practice at the time, if the translator wanted to disagree with the ideas being presented, barbarous footnotes were slipped into the text. In this instant, the footnote claimed that wheat prices in England could not be used to measure the general fertility of the Earth. Herschel responded by pointing out that the Moon raises tides of different heights in different places at different times, yet no one doubts that the Moon is responsible. So the wheat prices may be peculiar to England, but he was confident that other crops and other associations with sunspots would be found in other countries. In particular, he pointed out that an increase in the power of the Sun's rays might prove disastrous to a crop in a country that is "already fully hot enough for wheat."

If professional problems were not enough, Herschel also had concerns about John, his precious nine-year-old son. The boy was not strong, a circumstance the family blamed on having chosen an inadequate wet nurse. They had paid the woman to breast-feed John, a common practice at the time. Soon after the baby was weaned, the unfortunate woman declined and died, leaving the Herschels to worry about the inferior nourishment she had provided their only child.

John seemed little concerned by his fragility. He entered the prestigious boys' school of Eton, near the Herschels' home, but he did not stay long. One day on a ride through Eton, his mother was horrified to see him stripped and boxing with one of the larger pupils. She swiftly extracted her delicate child from Eton and placed him in a private school even closer to Slough. Then she arranged for John's education to be augmented with private mathematics tutoring.

As the pressures on Herschel mounted, he became dispirited about solar observing. Although he had perfected eyepieces to save his sight, he found that the telescopes themselves were suffering from the heat. If he used a mirror greater than nine inches in diameter, it warped under the heat, destroying the clarity of the image. Now approaching seventy, Herschel had no stomach for a new round of invention and let his solar studies lapse.

He began claiming extravagant expenses for a telescope he hardly used to help with income. The forty-foot-long telescope was the largest in the world. It hung within a wooden framework, itself over forty feet tall. It had been built with money from the King, who subsequently paid for its upkeep. The monarch's only stipulation was that he would sometimes bring guests to view the heavens through the mighty telescope. The trouble was, the telescope never worked well. It was too cumbersome to point accurately, and this meant that locating faint objects took too much time. Herschel soon returned to using a trusted twenty-foot-long telescope that blended power with practicality. His expense claims to the palace, however, showed that the dormant forty-foot accumulated a remarkable amount of wear and tear that needed repairing. In fact, the person who made the most use of the telescope was young John. He clambered over the frame as if it were a climbing frame.

In 1811, the mental strain finally took its toll. Herschel suffered a breakdown so severe that those around him feared his life was coming to an end. He pulled through, but the breakdown left him pale, shining only by the reflected light of past glories. His one hope was his son. The lavish attention bestowed on John was paying off. On 25 January 1813, the fragile William received word from Cambridge that his son had graduated as the top mathematics student of the year, a position known as senior wrangler.

William eased John's election to Fellowship of the Royal Society and began advising him on a career that would be both assured and allow plenty of time for scientific experimentation. To William, the choice was obvious. He began extolling the virtues of a life in the clergy. John vehemently opposed the suggestion, saying that he could not "help regarding the source of church emolument with an evil eye."

"The miserable tendency of such a sentiment, the injustice and the arrogance it expresses, are beyond my conception," replied William, who regarded the morality espoused by Christianity to be self-evident, regardless of whether one believed in God or not.

John wanted to train as a lawyer. William was so outraged at the prospect that he called the profession "crooked, torturous and precarious" and told John the mathematical studies at which he had clearly excelled were "of a superior kind."

The rift was patched over at Christmas, with William allowing John to enroll at Lincoln's Inn for legal training. The endeavor lasted just eighteen months. Unable to let go of his scientific interests, John exhausted himself trying to perform legal studies and scientific investigations. In 1815, on his doctor's recommendation, John abandoned Lincoln's Inn. He returned to Cambridge as a lowly mathematics tutor. Slowly, he embraced the life of a don and recovered his health.

At this, William again began to harbor aspirations of shaping John's career. In 1816, he took his son on a holiday to Dawlish, a fashionable coastal resort that was a favorite retreat of the romantic novelist Jane Austen. Old age was sapping William's strength faster than ever. As father and son breathed the restorative coastal air, William told of his unfinished astronomical work. Few, if any, professionals took his "sweeps" seriously enough to continue the endeavor. They seemed blind to the thousands of celestial clouds he had recorded. Called nebulas, he believed they were the vapors from which stars formed, but no one else seemed interested in exploring the nature of the nebulas, the stars and planets, or indeed the Sun. Instead, they slavishly continued to refine their star charts for use in navigation.

William's giant telescopes were the best in the world. Yet now they rotted unused in the gardens at Slough. The forty-foot telescope hung immobile, its mirror tarnished beyond repair. Its smaller brother, the twenty-foot telescope, was salvageable but was currently silent where once the noise of ropes threading pulleys spoke of it being pointed around the night sky. All seemed lost, unless John would continue his father's work.

John returned to Cambridge, his father's words weighing heavily on his conscience. By the autumn, he realized that there was only one noble course of action. If no one else would take his father's astronomical legacy seriously, then he had to. He sacrificed his own ambitions by resigning from Cambridge, moving back to Slough, and apprenticing himself to his father. Happily, his initial reticence swiftly transformed into a love of the awesome vistas that the twenty-foot telescope presented. He would often observe with a friend from Lincoln's Inn, James South. After one session sweeping their gaze across the mountains of the Moon and then the rings of Saturn, John wrote exuberantly that he believed large telescopes capable of showing anything.

When the aging telescope collapsed one night from John's renewed use, he immediately set about rebuilding it. In the process of re-creation, John glimpsed the man his recently knighted father had once been. He was amazed at the clarity of the patriarch's directions and realized that he still possessed "a mind unbroken by age." If only the same had been true of his stooped body.

With the telescope back in commission, James and John resumed taking turns at the eyepiece, hunting down the enigmatic clouds of nebulosity that littered the sky. They relished the complexity of each one they found. A particularly beautiful example drove James to blaspheme, "O good God! It is worth going to the devil for!"

John soon found others as passionate about astronomy, and together they discussed setting up an organization solely of astronomers, a crucible in which to indulge finely detailed discussions that lay beyond the scope of the Royal Society. Fourteen gentlemen astronomers drew up their plans on 12 January 1820 in the Freemason's Tavern, not far from John's lodging during his foiled attempt to study law. Among the supporters for the inauguration of the Astronomical Society was Edward Adolphus Seymour, the 11th Duke of Somerset and longtime Fellow of the Royal Society. He agreed to be the new society's president.

When news reached Sir Joseph Banks, who had presided over the Royal Society for forty years, he immediately sought out the duke. With considerable force Banks argued that separate bodies would fragment science, not only damaging the prestige of the Royal Society but also possibly threatening its very existence. Suitably convinced, the duke withdrew from the presidency of the Astronomical Society and resigned as a member too.

John asked his father to fill the vacancy. At first, Sir William declined citing his infirmities, but when Sir Joseph Banks died in the summer, Sir William changed his mind. His ever-dutiful sister, Caroline, wrote the letter of acceptance with the caveat that he be exempt from all duties. Sir William trembled uncontrollably; his mind was beset with worries about the forty-foot telescope, and he fretted endlessly

William Herschel's presidential portrait for the Royal Astronomical Society. (Image: Royal Astronomical Society)

about the safety of his old observing logs, fearing that his life's work would somehow be lost.

The melancholic gloom of Sir William's decline came to its crescendo eighteen months later when, in mid-August, the old man struggled for half an hour simply to stand upright. The servants returned him to bed, leaving his wife and sister to keep a desperate vigil at his bedside. John was touring the Netherlands at the time, one step ahead of the frantic letters calling him home. His intention was to visit the site of Napoleon's recent, definitive defeat. While inspecting the three farmhouses at the center of the battleground, John remained unaware that back in England, his father was facing his own Waterloo.

On 25 August 1822, after ten agonizing days trapped in the twilight, the eighty-four-year-old Sir William Herschel died.

THREE

The Magnetic Crusade, 1802–1839

Back in 1802, while William Herschel had been contemplating wheat prices and the possibility of natural climate change, a German naturalist by the name of Alexander von Humboldt had been looking down upon the wheat fields of the Cajamarca Valley in Peru, wondering what effect the spreading empires of mankind were having on the climate. He was in South America fulfilling an ambition that had begun in his youth, when he had listened to seafaring yarns and fallen in love with the idea of exploring the world to observe nature's beauty.

Approaching his thirtieth birthday, Humboldt set about transforming his ambitions into reality. He sailed to South America and found a largely unspoiled land of wonders in which he spent the next five years investigating and classifying everything he could, sometimes struggling for words grand enough to describe the volcanic landscapes he witnessed. Seeing such untamed vistas came at a price. He suffered plagues of mosquitoes and tropical diseases that took him to the brink of death. Underestimating his remarkable resilience, European newspapers reported him dead on three separate occasions; yet each time he recovered and marched on, deeper into the heart of the continent.

Humboldt's concerns about climate change began when he arrived at the crystal water of Lake Valencia in Venezuela, and discovered an anxious local community. Water levels were inexplicably dropping. Humboldt began investigating and discovered that forests help trap moisture-laden air, thus increasing rainfall over a particular area. Settlers around the lake had been busy felling trees to make way for homes and fields. As a result, not so much water was draining into the lake. Carrying with him the first concerns about the detrimental effects of deforestation, Humboldt continued his trek.

Throughout the journey, he employed guides and servants to carry his ever-growing specimen collection. Periodically he shipped this menagerie of live animals and cut stems back to Europe, in the expectation that he would eventually be reunited with them. Yet there was one discovery that he valued more than any other. It could be written on a single scrap of paper as a series of four numbers. It was the location of the Earth's magnetic equator.

The magnetic nature of the Earth has been obvious since ancient times when lodestones inspired the belief in magic powers. These naturally occurring rocks possessed the extraordinary ability to attract pieces of iron and, when suspended, would automatically point north. In the fifth century B.C. Lucretius wrote that the Greeks knew these wondrous rocks as Magnet, because they were found in Magnesia, Thessaly, in northern Greece.

In 1600, Queen Elizabeth I's physician, William Gilbert, extricated lodestones from the conjoined realms of mysticism and magic by performing repeatable, scientific experiments on them. He realized that lodestones could probably not cure gout or spasms, put one in favor with peers and princes, remove sorcery from women, put demons to flight, reconcile married couples, or even attract gold instead of iron after the lodestone had been pickled in the salt of a sucking fish. Nor would a lodestone lose its power after being anointed with garlic or placed near to a diamond. Instead, and rather less appealingly, all Gilbert found was that pairs of lodestones could be made to attract or repel each other, depending on how they were arranged. In other words, the magnets reacted with each other and that could mean only one thing. Since they always pointed north regardless of where they were located on the Earth, Gilbert correctly deduced that the Earth itself must be a gigantic magnet, possessing a north and south magnetic pole.

As magnetic compasses entered widespread seafaring use during the seventeenth century, a limitation in them became obvious: they did not point precisely to the north geographical pole (the point where the Earth's rotation axis exits from the planet in the northern hemisphere). So the north magnetic pole had to be located in a different place. Wherever a ship was positioned on the Earth, it would form a triangle with the two poles. Only when the ship sat on the same line of longitude as the magnetic pole, or 180 degrees away from it, would its compass point to geographic north. At all other points, there would be a variation that changed from one location to another. This variation, or magnetic declination as it came to be called, would need charting across the world so that correction factors to convert magnetic north to true north could be carried on all ships.

It was Edmond Halley of comet fame who, in the first years of the eighteenth century, weighed anchor at Deptford on the Thames estuary and sailed the Atlantic making the first large-scale chart of the declination. With Halley's tables as a reference, other problems soon became obvious. The declination itself was not steady but wandered on a daily basis, as if the magnetic pole itself was subtly changing position during the day before slinking back at night to start the same movement again the next morning.

Worse still was that yearly measurements of the declination of the naval yards in Greenwich, London, clearly showed that the north magnetic pole was drifting slowly every year. The movement was superimposed upon the daily advance and retreat of the declination and meant that Halley's correcting factors would need constant updating. This magnetic behavior confounded the natural philosophers. No other magnet varied its strength or direction of magnetism: clearly Earth is not the simple magnet that William Gilbert had described to the court of Queen Elizabeth I.

Halley concocted a theory in which the Earth was a hollow shell, filled with smaller spherical shells, like a set of Russian nesting dolls. All of the shells were magnetic and rotated at different speeds inside one another. The combination of their individual magnetic forces would then change with time. In a portrait of Halley as an eighty-year-old man, he can be seen holding a diagram of his hollow Earth theory. By the time of Humboldt, Halley's Byzantine solution had fallen from favor and natural philosophers were looking for something simpler. The Earth's magnetic field had three measurable components. The first was the strength, the second was the declination with its daily and yearly drifts, and the third was the angle that the magnetic field made to the ground: the inclination or dip.

When Humboldt left Europe in 1799, he carried an instrument capable of measuring this third element. The dip needle was a magnet suspended in such a way that it could swing up or down in response to the pull of the Earth's magnetic field. At the magnetic pole, the precise location of which was still unknown in Humboldt's time, the dip needle would point straight down into the Earth. What the German explorer became the first human to see was the dip needle resting parallel to the ground: indicating that he stood on the Earth's magnetic equator. The moment occurred at latitude 7 degrees 27 minutes south, longitude 81 degrees 8 minutes west, in the Peruvian Andes, 10,000 feet above sea level.

Despite being in the tropics, the plateau's elevation meant that cold, not heat, plagued the exploration party. Hailstones pounded them as they crossed the barren plains. Undeterred, Humboldt insisted that they stop and take magnetic readings. Carefully protecting the delicate instruments from the ravages of the weather, he watched in fascination as each time he took a reading, the dip needle flattened a little more. He also saw the first unmistakable evidence that the strength of the Earth's magnetic pull weakened as he drew closer to the magnetic equator. Pressing on, he found that the magnetic equator crossed the Earth's equator, running from southwest to northeast.

As he descended into the lap of civilization among the wheat fields of Cajamarca, Humboldt knew that these findings would be greeted with the utmost curiosity by the scientifically inclined in Europe. He went public with the results within a few months of his return home in 1804 and resolved to continue his magnetic studies. In May 1806, he rented a wooden potting shed outside Berlin and set up a precise compass. This time he wanted to investigate the unexplained daily movement and nightly retreat of the declination. With the help of an astronomer, Humboldt took magnetic declination readings every half hour, tracing the bizarre way the declination settled back to its starting point, before the arrival of the Sun at daybreak set it wandering again. What bizarre effect was the Sun having on the compass?

On 21 December 1806, something amazing happened. The magnetic needles went haywire, swinging through wild angles as if an earthquake were shaking the apparatus. Outside, Humboldt noticed that the aurora was lighting the sky, confirming Hoirter and Celsius's observations of over sixty years earlier that magnetic disturbances and the aurora went hand in hand. Struck by the invisible chaos, Humboldt coined the term "magnetischer Sturm" (magnetic storm) for such events.

Humboldt took the last of his six thousand magnetic observations in Berlin in June 1807 before relocating to Paris to begin the mammoth task of preparing his South American data for publication. He knew it would take years, but his initial estimate of just two years took ten times longer. He was slowed down partly by the huge volume of data he had collected and also because he became a household name as tales spread of his deprivations in the jungle, and of his bravery. Where once William Herschel had been the most famous scientist in Europe, now it was unquestionably Humboldt. It was even said that, apart from Napoleon himself, there was no one more famous in Europe than Humboldt.

King Frederick William III of Prussia made Humboldt a royal chamberlain, promising to waive all duties when Humboldt showed reluctance to accept. Nevertheless, with increasing frequency, Humboldt was called from Paris to attend court at Berlin. During these stays he argued that only Paris offered the necessary facilities for his work. In truth, he preferred the cosmopolitan lifestyle of the French capital.

However, when Charles X ascended to the French throne in 1824, a wave of ultraroyalism began to gradually squash the freethinking society that Humboldt loved. Political and military tensions were high between Prussia and France. In 1827 Humboldt was summoned not just to attend the Prussian court but also to join it permanently. Frederick William wrote, "You must by now have finished the publication of the works you considered could only be finished in Paris. I therefore cannot allow you to extend your stay any further in a country which every true Prussian should hate." Unable to deny the royal command, Humboldt returned to Berlin, settling into what he called the "nebulous atmosphere" of the king's domain. To distract himself from the court's parochial interests and its ceaseless oscillation between the palaces of Berlin and Potsdam, Humboldt turned again to magnetism. There was something he wanted to know about the magnetic storms: were they local, like rainstorms, or did they engulf the whole planet? Humboldt realized that he might be able to use his newfound fame and position at court to find out.

Also in Germany at around this time was Heinrich Schwabe, a thirty-six-year-old pharmacist. He lived in Dessau and had won the local lottery in 1825. His prize was a telescope, through which he began observing at once, having studied astronomy alongside pharmacology and botany in Berlin. Schwabe was the eldest of eleven children. In the daylight hours, he toiled in the pharmacy that had once belonged to his grandfather, selling potions and poultices to provide an income for his widowed mother and her dependent children. At night, he discovered the wonders of the celestial realm, becoming ever more fixated on astronomical discovery.

Less than a year after winning the telescope, he diverted money away from his family to order a better instrument from Joseph von Fraunhofer in Munich. He also canvassed other astronomers for advice on valuable observing projects. K. L. Harding of Göttingen suggested he concentrate on the capricious solar spots. As an added incentive, Harding mentioned that the reward of such scrutiny might be the discovery of a hypothesized planet inside Mercury's orbit.[1]

On 30 October 1825, Schwabe turned his telescope to the Sun and began a series of sunspot observations that he would continue for the next forty-two years. Every clement day, Schwabe noted down the positions and descriptions of the sunspots. The observing log for his first tentative year alone ran to sixty pages.

During 1829, probably because his siblings were now largely independent, Schwabe sold the pharmacy and devoted himself to astronomy full time, little knowing that he was on course for a major discovery that would put the study of sunspots on a direct collision course with the study of terrestrial magnetism.

To investigate the extent of the enigmatic magnetic storms, Alexander von Humboldt needed a string of stations around the globe, each equipped with instruments to measure Earth's restless magnetic field. He built the first in Berlin, housed in a wooden cabin in the garden of Abraham Mendelssohn-Barthody, father of the famous composer. Then, he

[1] Mercury was not following the orbit prescribed for it by Newton's law of gravity. Many thought another planet, which they named Vulcan, must be tugging on it through gravity. Many searches were mounted but no planet was ever found. Einstein's general relativity theory finally explained Mercury's peculiar motion.

used his reputation to arrange for similar readings to be taken in Paris and organized a conference in Berlin, tempting the reclusive physicist Carl Friedrich Gauss to attend. Gauss possessed considerable talent, but he was notorious for keeping his discoveries private, sharing only the ripest fruits of his labor. He took but few students and treated most of his associates with lofty arrogance. Those closest to him blamed this lack of manners on having never truly recovered from the death of his first wife, Johanna Osthoff, in 1809, when she was just twenty-nine and he was thirty-two. But if anyone could help understand the restless nature of the Earth's magnetic field, it was Gauss.

Humboldt lavished attention on him and, by the end of the conference, had persuaded Gauss to contribute to the pan-European effort. Gauss returned to Göttingen and set about developing new instruments. He even began working with a colleague, Wilhelm Weber, in the endeavor. Having Gauss's brainpower on side was an academic coup, but if Humboldt were to succeed in his goal of a global network, he would need to extend it beyond Europe.[2] This is where he used his position in the Prussian court. On a successful diplomatic visit to Russia in 1829, Humboldt arranged for stations to be built across Siberia and into Alaska, which was then under Russian jurisdiction. Next, he needed the help of the British and their world-spanning empire. Although America had declared its independence in 1776, the British Empire still held a swathe of North America in the form of Canada. Across the Atlantic, the empire swept through Africa and the Indian subcontinent to Singapore and came to rest in the trio of antipodean landmasses: Australia, New Zealand, and Van Diemen's Land (Tasmania).

In Britain, curiosity in terrestrial magnetism was running high, with John Herschel close to the center of interest. The undeniable coincidence of magnetic storms with auroras led Herschel to believe that both were some form of meteorological phenomenon. With increasingly sophisticated magnetic instruments providing better windows onto this invisible world, Herschel's peers agreed that a string of new observatories was urgently needed.

[2] Upon Gauss's death, in 1855, his brain was preserved and studied. It was used to advance phrenology, a discredited science that sought to ascribe personality traits to the development of different areas of the brain. Gauss's brain featured a number of pronounced regions and these were thought to be the seat of his genius.

Most believed that the magnetic storms should be the focus because of their singular nature. To miss one was to lose it forever. One man disagreed. Colonel Edward Sabine had been sent by the Royal Society to accompany John Ross's 1818 expedition to find the Northwest Passage, and again on William Edward Parry's great Arctic expedition of 1819–1820, in part to carry out magnetic investigations. He was adamant that the day-to-day variations were just as important because they could reveal the underlying state of the Earth's magnetic field.

Sabine was a quick-witted man with an innate sense of politics and began lobbying both the Admiralty and the Royal Society for a great voyage of physical discovery in the southern hemisphere. He pointed out that no one had taken magnetic readings from the great southern oceans. Although John Ross and his nephew James Clark Ross had recently discovered the location of the north magnetic pole, no one had a clue to the whereabouts of its southern counterpart. Despite quibbling over storms versus day-to-day readings, Herschel backed Sabine's proposal, but, before they could capitalize on their growing momentum, a rift in the British scientific community cast them on opposing sides.

It was 1831. John was married with two children. He was president of what was now the Royal Astronomical Society, recently granted a royal charter by King William IV. Sir John had recently been knighted for services to science, an honor his father had waited until age eighty to receive, but the restless thirty-nine-year-old wanted more. Always the activist, he and others considered the Royal Society old fashioned and conceived a plan to reform it from within. The first step was proposing John for its president, which a co-conspirator duly did. John's opponent was His Royal Highness, the Duke of Sussex. In the ensuing battle, the Royal Society's Fellows split along the lines of tradition over reform. Sabine sided with the traditionalists. In the final vote, John lost by a narrow margin.

A number of the defeated reformers founded the British Association for the Advancement of Science (BAAS), in open rivalry to the Royal Society.[3] The BAAS would meet every year in a different provincial

[3] For a fuller discussion of the BAAS's formation, see Roy M. MacLeod and Peter Collins's 1981 book, *The Parliament of Science: The British Association for the Advancement of Science, 1831–1981,* published by Science Reviews Ltd.

John Herschel retreated to South Africa and extended his late father's catalog to the southern skies. John constructed the twenty-foot telescope at Feldhausen. (Image: Royal Astronomical Society)

city so that the latest science could be presented and discussed away from the stifling elitism of London. Sabine snubbed the new organization. Herschel lent his support but, embarrassed by his Royal Society defeat, began preparing to leave England. His plan was to establish himself in South Africa to perform the first astronomical sweeps of the southern skies. In this way he would bring his late father's work to glorious culmination. When John's mother died he hastened his plans, and, on 13 November 1833, he loaded his wife, his children, now numbering three, and his twenty-foot-long telescope on a ship at Portsmouth and sailed for the Cape of Good Hope.

With John abroad and many potential allies now working from within the BAAS, Sabine was isolated. In Germany, Gauss announced a breakthrough. He had succeeded in developing better magnetic instruments and was putting together his own network of magnetic observatories that ran the length and breadth of Germany. Norwegian and French scientists were also becoming increasingly active. The Norwegian parliament had even funded a geomagnetic expedition in preference to a royal request for money to build a new palace.

At the 1834 BAAS meeting in Edinburgh, scientists voiced their indignation at the British government's seeming reticence to fund magnetic research. They formed a subcommittee charged with persuading Parliament to make magnetic equipment available throughout the British Empire. Watching from the sidelines, Sabine realized this could usurp his proposed voyage of discovery. In a political quickstep, he joined the BAAS and won a place on the committee.

Once in, he began to inveigle his southern voyage into the concept of colonial observatories. As a result, the scope of the project grew until Sabine was masterminding the greatest scientific undertaking the world had ever seen. He succeeded in interesting the Royal Engineers in manning the observatories, arguing that compasses were as much a military aid as a navigational one, and, at the 1837 BAAS meeting, he stoked the patriotic fire within anyone who would listen. Was Britain to let its once prominent lead be overtaken by the Germans or the French, through a combination of indifference and niggles over cost?

This public outrage of Sabine's was never more than grist for the mill of his ambition. In private, Sabine colluded with Humboldt over how best to achieve a network that was to both their advantages. With one hand, Sabine spouted his rabble-rousing sentiments at the BAAS. With the other, he orchestrated dignified written requests for British assistance from Humboldt to appear at strategic times on the desk of the Royal Society's president, the Duke of Sussex.

Thanks to his lionhearted advocacy of a magnetic crusade, sustained for years on end, the government was edging toward an agreement. All Sabine needed was for someone eminent to give it a final seal of approval. That person returned to England on 11 May 1838.

After five years, John Herschel was back with a family that had doubled to six children, to whom he referred as "the Little Bodies." His achievements at the Cape made even his father's early surveying look leisurely. Word of nearly five thousand newly discovered double stars, nebulas, and star clusters preceded him, and, wherever John went, he found himself held in higher esteem than ever before. His southern accomplishments banished forever the embarrassment of his failed bid for the Royal Society. One eminent Fellow, geologist Charles Lyell, wrote with an ironic pen, "Fancy exchanging Herschel at the Cape for Herschel as President of the Royal Society. . . . I voted for him too! I hope to be forgiven for that." Just six weeks after Herschel's return, Queen Victoria created him a baronet at her coronation.

The BAAS requested he speak in favor of the magnetic crusade at that year's meeting, in Newcastle. Herschel agreed but added his own twist by insisting that the role of the observatories be extended to meteorological work too, since both air pressure and temperature could affect compass readings.

After the public success of the Newcastle meeting, Herschel raised the subject in private, over dinner with the Queen and the prime minister, Lord Melbourne. Sabine continued to discuss practicalities with his contacts at the Admiralty, and eventually, on 11 March 1839, the Antarctic expedition was given the go-ahead. The following year, magnetic stations were founded at Greenwich, Britain; Dublin, Ireland; Toronto, Canada; St. Helena in the South Atlantic; Cape of Good Hope, South Africa; Van Diemen's Land; Madras, Simla, and Bombay in India, and in Singapore. Each would be funded for three years at a cost of £2,000 each.

The magnetic crusade had begun.

FOUR

The Solar Lockstep, 1839–1852

On New Year's Eve 1839, John Herschel solemnly gathered his wife, family (to which he had recently added another child), and housekeeper in the garden at Slough. They were to lay his father's largest telescope, the forty-foot, to rest. With the completion of the southern sky catalog, John considered his promise, to complete his father's work, fulfilled. The long nights under the African stars had left him with rheumatism, and he was eager to abandon observational astronomy and pursue his interest in the recently invented techniques of photography.

He ushered the gathering inside the telescope's six-foot-wide tube, now lying prone on the grass, and led them in an eight-verse requiem he had composed to honor the telescope. Workmen were subsequently employed to seal the tube and dismantle the towering wooden structure that had supported the telescope and doubled as John's childhood climbing frame.

In England, and indeed throughout Europe, the accepted theory of the Sun was still the one espoused by his father, William, over thirty years previously. John engineered this by publishing his father's theory in the *Treatise on Astronomy,* a book he subsequently updated in 1849 and renamed *Outlines of Astronomy*. Although he rephrased it in the scientific language of the day and prudently omitted all references to the Sun's inhabitants, it was essentially his father's solar theory: the visible surface of the Sun was a great atmosphere surrounding a solid planetary body. What William had described as looking like orange peel, John interpreted as flocculent atmospheric markings that were thought to be luminous clouds intermingled with a transparent gas. He went on to describe

Even after abandoning observational astronomy, John Herschel was captivated by the discovery that magnetic disturbances on Earth walked in lockstep with the number of sunspots. (Image: Royal Astronomical Society)

the sunspots as openings, possibly tornados, that allowed sight of the Sun's dark surface beneath.

To be fair, the text was not all recycled from his father; John had devised a method of measuring the strength of sunlight, something his father had only dreamed about. The lack of such readings three decades before had forced William to use wheat prices as a proxy for the Sun's outpouring of heat. At the time he had written of the need for a device capable of measuring an absolute quantity of light, just as a scale balance and a set of measures could weigh meat.

John invented it and called it an actinometer. It was essentially a bulb of water that was exposed to sunlight for a set amount of time. Afterwards, the temperature of the water was taken and the rise used to calculate the energy received from the Sun. It could be used for daily comparisons but failed to take into account changes in atmospheric conditions, such as cloud cover. While at the Cape of Good Hope, he took regular readings with the actinometer. He also performed some rather more eccentric experiments such as the day he placed a fresh egg in a tin cup and then lay a pane of glass across the top. Returning with his wife and six children some time later, he retrieved the cooked egg, burning his fingers in the process. Ceremoniously, he cut the egg into pieces and doled it out, so that all could say they had eaten an egg boiled hard by the South African Sun. Suitably impressed by his newly found culinary skills, a week later he cooked a mutton chop and potatoes the same way. "It was thoroughly done, and very good," he recorded in his diary.

Both the *Treatise on Astronomy* and *Outlines of Astronomy* were hugely successful. Herschel revised *Outlines* twice, and it became probably the most influential astronomy text of the nineteenth century, certainly a standard text for students. One young person who would have been richly exposed to its influence was Richard Carrington, the man destined to witness the solar flare of 1859. The son of a brewer from Brentford, Middlesex, Carrington entered Trinity College, Cambridge, to read mathematics during 1844. His father had been preparing him for a career in the Church, even sending him to live with a clergyman by the name of Blogard. However, Carrington's passion was for mechanical technology and, once at Cambridge, he indulged his natural aptitude by learning how to use scientific instruments to transform nature into numbers. The logic of mathematics could then give new insight into the real world. This concept was confirmed in dramatic fashion when, as Carrington commenced his third year, the planet Neptune was discovered. Remarkably, the world was found not with a telescope but with pen, paper, and mathematics.

Astronomers had been tracking Uranus for over fifty years. At every turn the stubborn planet had confounded them by deviating from its assigned orbit. They reasoned that another equally mighty planet had to be out there, pulling Uranus off-course with its gravity. Astronomers in both England and France began trying to calculate the position of this unseen orb. In England, John Couch Adams, a shy and overworked Fellow at St. Johns College, Cambridge, took up the challenge. His opposite number at the Paris Observatory was the dictatorial Urbain Le Verrier, of whom it was once written, "I do not know whether M. Le Verrier is actually the most detestable man in France, but I am quite certain that he is the most detested."

Both men struggled with the numbers until they deduced the whereabouts of the suspected eighth planet. Adams sent his prediction to the Royal Observatory at Greenwich. It was sent back to Cambridge to become the subject of an unsuccessful search by the director of the university's observatory, Professor James Challis.

In contrast, Le Verrier published his prediction. He subsequently wrote to Johann Gottfried Galle at the Berlin Observatory. The German happened to have a recently completed star map for the part of the sky where Le Verrier thought the planet would be. Galle began searching that very night. Peering through the eyepiece, he called out the positions of the stars and his student, Heinrich Lugwig d'Arrest, checked them against the star charts. Less than an hour after they began, Galle described a faint star. "That star is not on the map," exclaimed d'Arrest. They had found Neptune lying almost exactly where Le Verrier had predicted.

The discovery caused both a sensation and a scandal. It was sensational because mathematics had endowed astronomers with scientific foresight. Neptune had been discovered not in an observatory but on the page; the telescope merely confirmed the validity of the mathematics. From now on, discussion would take place solely based upon mathematically measurable facts. The scandal was that the English astronomers could have been first.

A reanalysis of the Cambridge observing logs showed that Challis had seen Neptune using Couch Adams's prediction but had mistaken it for a star. Although others were complicit in the failure, Challis became the scapegoat. His professionalism was called into question and he was publicly humiliated for his mistake. Despite this, Richard Carrington saw a different facet of the disgraced professor, who remained an engaging public speaker. It was during Challis's lectures that Carrington resolved to pursue astronomy rather than enter the Church. Carrington wrote that he was "more naturally adapted for the pursuit of some physical science involving observation and mechanical ingenuity than for the public exposition of the doctrines of a body with which I have ever had little sympathy, much as I esteem individuals belonging to it." Perhaps surprisingly, given the upbringing Carrington had received, his father blessed this change of tack.

Upon graduation in 1848, the young man drew up his plans. His ambition was to own a world-class observatory: one with sturdily mounted telescopes that contained the finest optics, clocks that stuck to rigid time, and precision ancillary equipment that could be used to measure the positions and sizes of astronomical objects. With a portion of the family wealth at his disposal, Carrington could have begun building straight away but he believed that his relative inexperience would lead him into "wasteful and injudicious expense." So he first set about finding himself a position of paid employment within an existing observatory, to learn all he could before striking out on his own.

It was no easy task; the Universities of Cambridge and Oxford both had observatories but no vacancies. The Royal Observatory at Greenwich similarly had no openings. By chance, the University of Durham had founded an observatory a decade before and now required a new observer. Carrington applied and became the fourth person to hold the post.

Durham's observatory consisted of a series of domes paid for by private donations in which were housed telescopes and other equipment that had unexpectedly been offered for sale by an amateur astronomer who had no further use for them. The Reverend Temple Chevallier, Durham's professor of mathematics, administered the running of the observatory and believed that the proof of divine wisdom would be found in the study of astronomy. Presumably Carrington kept his own views on religion quiet at the interview.

Carrington's living quarters were situated in the observatory itself, and he used this proximity to its best advantage, quickly learning the art of observing. Night after night he set the telescopes on the sky. Mostly he calculated the orbits of asteroids and comets, filling reams of paper with his correction factors and mathematical calculations. He found the clock too far away from the telescopes to be able to reliably hear the tick-tock of the escapement. Needing this to time his observations, he installed a hearing tube to funnel the sound to him at the eyepiece. He was also dissatisfied with the stated value of Durham's longitude, an essential value for turning his observational readings into accurate positions. The zero line of longitude is defined to run between the north and south poles, passing through the Royal Observatory at Greenwich en route. It is known as the Prime Meridian, although it would not be internationally recognized as such until 1884. The Durham observatory, while lying close to this meridian, is not exactly on it, and so a correction factor has to be applied to make the telescope's readings match those of Greenwich. Carrington was suspicious of the estimated correction factor and so devised a way to precisely measure the Durham longitude.

He obtained three clocks in sturdy box frames and set each of them to midday, judged by when the Sun reached its highest position in the Durham sky. Clocks in Greenwich would be set in a similar fashion but, because the observatories are at slightly different longitudes, the clocks will be a few minutes different from those in Durham. To calculate Durham's longitude, all Carrington needed to do was compare the clocks set at Durham to those set at Greenwich. That meant taking the clocks on a journey. He loaded them onto a train and nursemaided them to London. Even though the clocks had been designed to withstand the rolling waves at sea, Carrington took great care not to jolt them unnecessarily. When he arrived in London, he chartered a sprung carriage and had it driven a circuitous route to Greenwich, avoiding the worst of the cobbled streets. Once arrived, he found the Greenwich chronometers to be some 6 minutes 19 seconds ahead, allowing him to calculate that the observatory in Durham lay 69 miles west of the prime meridian.

There were other problems that he could not fix so easily. When comparing his observations of asteroids with those taken by observers from Oxford and Cambridge, he found that he lost sight of faint objects much sooner than his peers. This frustrated him and he blamed it on the poor quality of the telescopes he had at his disposal. Nevertheless, Carrington was experiencing a meteoric rise within astronomical circles. His observations were regularly printed in the pages of *Monthly Notices of the Royal Astronomical Society* as well as in *Astronomische Nachrichten,* and he became a Fellow of the Royal Astronomical Society.

In tandem with nocturnal observations, Carrington began observing the Sun. He charted the positions of sunspots and acquainted himself with the various forms these markings could take. He also began preparing for an expedition to see the total solar eclipse of 1851. It must have been an exciting prospect. Total eclipses were highly prized for they revealed the ghostly outer atmosphere of the Sun. This tenuous veil was usually hidden by the Sun's glare, and astronomers relished the few minutes that a total eclipse afforded them for witnessing its strange beauty. Good fortune had dictated that this particular eclipse would be visible from nearby Sweden.

On 19 July, Carrington set sail on the courier *Steampacket*, bound for Göteborg, Sweden. From there he teamed up with G. P. Bond, an astronomer from Harvard College Observatory in Cambridge, Massachusetts, and they took the steamboat to the village of Lilla Edet. They arrived three days before the eclipse and began scouting locations. On the morning of the event, they were dismayed to see the sky overcast and drizzling with showers and decided to split up, hopefully doubling their chances of one seeing the eclipse. Bond stayed close to the village while Carrington headed for a rocky outcropping some two miles away, close to the local canal he had noted during his previous days' reconnoiter.

Despite the miserable weather, once in position Carrington set about his preparations. The task aroused the interest of the local canal overseer who, upon learning what was to happen in three hours, swiftly detailed some of his workmen to assist in the erection of the tent and observing equipment. While in the throes of their endeavor, the weather began to change, and by the time the allotted moment had arrived everything was set and the sky was clear.

Carrington placed his right eye to the telescope and watched the Moon begin its obstruction of the Sun. He sketched the positions and appearances of the sunspots in the hour it took the Moon to slip completely in front of the Sun. All seemed normal until the last five minutes, when an eerie twilight crept upon the land. This was not the rosy glow of dusk but a disturbing gray light. It suddenly grew cold and then, in the blink of an eye, all light from the Sun was extinguished and its ethereal crown shimmered into view. Stretching outward were pale streamers many times the diameter of the Sun. Carrington had mere minutes to sketch the features.

Much closer to the Moon's black silhouette, Carrington saw four tongues of pink flame curling upward from the hidden solar surface. They seemed to hang there, motionless. He swapped observing eyes to make certain they were real and not the product of a defect in his right eye. Sure enough, the flames remained. He sketched the scene until the Sun's brilliance was restored, just a few minutes later, when the Moon slipped off the Sun. Carrington hurried to find other observers and compare notes. He collected these in written form and translated them with a Swedish dictionary that he described as "very imperfect." The quality of the translation not withstanding, others confirmed that they too had seen the tongues of pink flame.[1]

What could cause such eruptions? he wondered. His only clue was that they had appeared to line up with the sunspots—could there be some connection? With his mind full of such intriguing thoughts he returned to Durham and found a special job waiting.

Impressed with his skills, the warden and syndicate responsible for the observatory asked him to report on the facility's present state. Carrington did not hold back. He lambasted almost every aspect of the observatory and its management, beginning with a criticism of the way it had been founded. "In the establishment of an observatory it should first be decided what the course of observations is to be, and then find the requisite instruments—not that an observatory and instruments be got together and then enquiry be made to what they can be applied," he wrote.

As for the equipment: the general purpose telescope, a seven-foot-long instrument made by one of the world's leading telescope manufacturers, Fraunhofer, was "not equal to the reputation of its maker." The mounting, even though Carrington had strengthened it with wooden poles, displayed more "the nature of a blancmange than of a rock." The transit circle, the telescope used for determining the time at which stars

[1] This was not the first time such prominences had been observed. Birger Wassenius saw some at the eclipse of 1733, coincidentally from near Göteborg.

reached their highest altitude, was unreliable and sometimes showed rings around the stars and "little fictitious companions," indicating that stray light was entering the tube somehow.

According to Carrington, most of the equipment should be disposed of "at any sacrifice." He had done everything he could to improve the situation with petty cash and clever thinking, now it was time for a substantial investment in newer equipment. He offered the syndicate a £50 donation and a further £1,000 of his family wealth to achieve this refurbishment. The higher figure would be in the form of a mortgage, repayable to Carrington over ten years at "a moderate rate of interest."

The twenty-five-year-old observer wrote separately to Archdeacon Thorpe, chairman of the syndicate, stating that his growing reputation would not allow him to work with such inferior instruments for much longer. Also, his personal position as an independently wealthy gentleman rendered him unwilling to continue acting as a subordinate. He could no longer work with the Reverend Chevallier on a cordial basis. Either the observer's post was to be removed from the direction of the chair of mathematics and provided with better telescopes or, Carrington warned, he would be forced to resign and establish his own private observatory.

In November, the syndicate called Carrington before them to discuss his report, and again three weeks later. After the first meeting, Carrington thought they were willing to entertain his ideas, but the actual proposals that were made to him at the second fell far short of his expectations. He referred to them as shabby and tendered his resignation. Finally leaving Durham behind in 1852, he began the painstaking search for a site on which to found his own observatory.

Almost as soon as Colonel Sabine's magnetic crusade had launched, the cabal of scientists behind the endeavor began to fragment. Sabine's main role had been to oversee the southern voyage, but now, as morale dropped, he used his army connections to commandeer the organization of the land-based observatories, too. He established a battery of mathematicians, known as computers, at Woolwich Arsenal in London and began to hungrily collect the data from the various stations around the globe. He also made a move on the Royal Observatory at Greenwich, seat of the formidable Astronomer Royal, George Biddell Airy.

While lobbying for the magnetic crusade, Sabine had prompted Humboldt to write a number of letters to the Royal Society. These were more than mere notes; they were dossiers of why magnetic observations were important and how a global network could be achieved. The Royal Society asked Airy to scrutinize the proposals. Airy agreed that such observations were necessary but deplored the scale of the proposed British network. He argued instead that such observations fell within King Charles II's 1675 remit for Greenwich, which had been to advance navigation. Airy therefore forwarded a modest proposal in 1837, two years before the crusades launched, to establish a magnetic observatory at Greenwich. He offered his own time to superintend the project freely but needed more staff to operate the instruments.

Airy believed in efficiency above all else and, as a result, organized his staff down to the last detail. When Airy needed staff members to attend an experiment to measure the Earth's gravitational field from inside Harton Colliery in county Durham, he drew up detailed plans for their journey, including which trains to catch, and where to change, so that they would not get lost. He even packed soap and towels in with the scientific instruments, lest they forgot personal hygiene while out of his immediate sight.

Airy argued that manning the new magnetic instruments would overextend his staff. The government disagreed, granting Airy only the funding necessary to build a wooden pavilion and purchase the necessary equipment so that Greenwich could start feeding results to Humboldt and Gauss in Germany. The pavilion was cruciform in shape and stood on concrete foundations. It was constructed of wood and held together by bamboo pegs to remove the magnetic influence that nails might have introduced. It was completed in 1840 and the readings began. On most days they were taken every two hours, but there were also special "term-days," decreed by Humboldt, when readings had to be taken every five minutes, so as to provide a nearly continuous profile of the Earth's magnetic behavior.

As Airy feared, fitting in the magnetic readings soon became a burden to his staff. Upon hearing this, Sabine jumped straight in with offers of help. Exasperated with the colonel's appetite for data, Airy refused.

Unable to extend his dominion over Greenwich, Sabine saw an opportunity to overshadow its authority. King George III's observatory at Kew was lying empty. Sabine proposed using it as a central laboratory for physics and meteorology, entirely separate from Greenwich. Airy was outraged and, forgetting the strain that the magnetic readings were placing on his staff, argued that the success of the new magnetic pavilion proved that Greenwich could be extended to fulfill whatever role a putative national physical laboratory required. The Royal Society deliberated over Sabine's plan but, on the advice of John Herschel, who was also growing weary of Sabine's rapacious appetite for new establishments, turned down the opportunity to transform Kew.

Never outsmarted for long, Sabine approached the burgeoning British Association for the Advancement of Science with the same proposal and met with a much warmer response. They acquired Kew in 1842, and Sabine set about transforming the place into the British center for geophysical research. He also won a further six years of funding for the magnetic observatories in the imperial lands.

By the early 1850s, he was swimming in data. His staff collected, tabulated, and plotted every conceivable permutation. Herschel thought the endeavor obsessive, calling it "chartism." Nevertheless, it paid off. One of Sabine's charts recorded the number of magnetic storms that took place each year, another the average daily variation of the compass needle. Both seesawed in a similar fashion, meaning that the years with the greatest number of magnetic storms were also those of the greatest daily variation in the three magnetic components. But the truly astonishing thing was that Sabine was about to see this shape of graph again—and not in magnetic data.

Sabine's wife was translating Alexander von Humboldt's epic work, *Kosmos*. It was the summation of the German naturalist's lifework, drawing together as much information about the Earth and its place in the Universe as possible. In the pages of the third volume, he drew attention to an overlooked set of sunspot observations that showed something amazing.

Heinrich Schwabe, the pharmacist-turned-astronomer from Dessau, had looked out from the makeshift observatory in his attic every clear day since 1825 and counted sunspots. During some years there were so many that he became confused in the attempt to record them all. It had been like this in 1828, shortly after he began keeping records, and again in 1837. In the intervening years, the spot numbers had first fallen and then risen again. By 1843, he had accumulated enough data to see the pattern repeating itself.

He concluded that the number of sunspots waxed and waned on a roughly decade-long cycle. He predicted that the next sunspot maximum would occur around 1849, followed by a minimum about five years later. He published these ideas in the German astronomy journal, *Astronomische Nachrichten*. During each subsequent year, he published his annual tally of the sunspot number, gradually showing that his prediction was coming true.

This was a major leap forward. Until now the spots had remained stubbornly unpredictable. A repeating pattern might be exactly what was needed to understand their origin, but no one seemed to recognize the importance of the results except Humboldt. He published Schwabe's table, fully updated to 1850, in *Kosmos*, ensuring that it would be widely read.

When Sabine saw the table, he recognized the underlying pattern immediately. He compared the variation of the sunspot numbers with that of magnetic storms and found that they moved in lockstep with one another. He compared the solar cycle with the average daily variations in the magnetic components. They, too, were somehow tied together. It must have been a dizzying realization. When the numbers of sunspots were high, the disturbances of the Earth's compass needles were greatest, as were the chances of a magnetic storm on Earth.

As his wife sent her translation to the publisher, Sabine hastened to alert the Royal Society of the sunspot-magnetic storm connection. He immediately wrote and dispatched a paper on the subject. While waiting for it to endure the Royal Society's vetting procedure prior to being read at a meeting, he received a letter from John Herschel. Past squabbles with Sabine forgotten, Herschel had just received a published copy of *Kosmos* and had noticed Humboldt's reference to the sunspot cycle. Admitting it was new to him, Herschel called it extremely curious and asked, "What can there be to determine a periodicity of this kind on the Sun?"

Sabine could hardly contain himself. He wrote back by return post, stating, "With reference to Schwabe's period of 10 years having a minimum in 1843 and a maximum in 1848, it happens by a most curious coincidence (if it be nothing more than a *coincidence*) that . . . I trace the very same years." He then described his similarly varying magnetic data.

Herschel was gripped by the importance of Schwabe and Sabine's correlation. The Earth was clearly in thrall of some extraterrestrial force, probably magnetic and probably coming from the Sun. Did this mean that the Sun was a magnet? Were the sunspots caused by magnetism? And if so, how could they possibly prove it? How could that same magnetism reach across space to seize the Earth?

Herschel wrote to Michael Faraday, the vaunted experimentalist who was investigating the links between electricity and magnetism, saying, "We stand on the verge of a vast cosmical discovery such as nothing hitherto imagined can compare with."

FIVE

The Day and Night Observatory, 1852–1858

Carrington's search for a site on which to build his dream observatory took three months. In June 1852, he chose Furze Hill, a leasehold plot on a rural estate in Surrey. The nearest conurbation was Redhill, which was a stop on the London-to-Brighton railway line, allowing him easy access to the capital.

Financed by the family brewery, he engaged builders to begin construction at once. Not only was this to be his observatory, but he was going to make it his home as well. He supervised the construction of a three-story manor house with bow windows and a high-pitched roof. The house faced south, and to the east he constructed the observatory wing. It doubled the width of the house and possessed a tall dome on its far end. By the end of July, the construction work was substantially complete and Carrington began to equip the observatory. He ordered a state-of-the-art telescope with the best four-and-a-half-inch diameter lens available, handmade by one of Britain's finest telescope manufacturers, Troughton and Simms. This was the "equatorial" telescope, so-called because it could swivel parallel to the Earth's equator as well as up and down. Thus it could point anywhere in the night sky. It would be housed beneath the dome.

Halfway along the observing wing was a shuttered slit in the roof, as if some giant had taking a slice out of the building. This is where the "transit circle" or meridian telescope would be housed. Also ordered from Troughton and Simms, it was used by Carrington to take accurate positions of stars as they crossed the meridian, the imaginary line that ran north-to-south through the house. Unlike the versatile equatorial,

Richard Carrington built this manor house with an adjoining observatory at Redhill, Surrey. It was from here that he witnessed the solar flare. (Image: Royal Astronomical Society)

the meridian telescope would be fixed to point only toward the meridian, although it could move up and down to reach stars at different altitudes.

With the problems of the wobbly transit circle at Durham still fresh in his mind, Carrington supervised every step of the building process. The foundations for the telescope's support pillars were excavated to a depth of five feet, into which stone and concrete were packed. A single five-inch-thick flagstone was laid on top and precisely leveled. Once satisfied, he allowed the pillars to be constructed. They rose like megalithic standing stones, three feet apart so that the five-foot-long meridian telescope could be securely pivoted in between. Nearby, he added a scientifically accurate clock.

Carrington also built quarters for a live-in assistant at the observatory and employed George Harvey Simmonds. A housekeeper was quartered in the manor house's third-story attic rooms. With the instruments and staff in place, Carrington moved on to the painstaking task of precisely adjusting the observatory and its equipment. If he was going to produce anything useful, he had to understand the exact characteristics of the telescopes, so that he could correct their undoubted quirks of manufacture. Once everything was calibrated, he planned to begin compiling a star catalog of the northernmost regions of the sky, areas that had been neglected by the larger observatories.

While he was preparing the observatory, news of Sabine's correlation between magnetic disturbances and Schwabe's sunspot cycle swept through the astronomical community. So too did the work of a Swiss astronomer called Johann Rudolph Wolf. Working at the University of Bern, Wolf collated the sunspot records of previous astronomers in an attempt to extend Schwabe's cycles back to the time of Galileo's first telescopic sunspot observations. He managed to find evidence of Schwabe's cycle all the way back to 1755, allowing him to revise Schwabe's rough decadal estimate to 11.11 years as the average length of a cycle. Before 1755, however, the observations all but petered out. Sunspots, it seemed, had been rare in those years—or the observation of the Sun had simply been neglected.

The 11.11-year cycle seized Carrington's attention. His most profound belief was that the Universe was founded on logical principles, not capricious whim. On days when he was not adjusting the telescope, he traveled to London to make his own study of the solar observations held in the library of the Royal Astronomical Society. He became swiftly disillusioned with the efforts of previous astronomers, finding that the observations had been undertaken with an almost careless disregard for positional accuracy. Different observers seemed to choose different values for the Sun's rotational period. Neither did they seem to agree on the tilt of the Sun's equator and rotation axis. Even when Carrington found a good set of observations, he was appalled at the casual way the observer sometimes made observations and sometimes failed to bother.

Carrington therefore resolved to use his new observatory not only for the northern star catalog but also for a thorough study of sunspots. This would fully occupy his time, allowing him to engage in the methodical collection of stellar positions at night while exercising his imagination during the day as he searched for meaning in the sunspots.

On 9 November 1853, he cranked open the shutter on the dome and swung the equatorial telescope in the direction of the Sun. Instead of looking through the eyepiece, he carefully placed a distempered board in position, so that the Sun's bright image fell onto the makeshift screen. With a pair of gold crosshairs placed in the eyepiece to cast diagonal shadows across the image, he began sketching what he saw. As the Earth turned, the Sun's image drifted across the field of view. He timed how long it took each sunspot to pass across the static crosshairs so that later he could use geometry to translate the timings into solar latitudes and longitudes. Not content with doing this just once, he repeated the timings at least twice so that he could average the results for better precision. He was determined that this study would be thorough. He would display the same fortitude as Schwabe, recording the appearances and positions of the sunspots every clear day for the next eleven years, enough time to see the Schwabe sunspot cycle for himself.

During 1854, with the transit circle finally adjusted to perfection, Carrington and Simmonds began the catalog of the northern stars. When not observing, they devoted themselves to the grueling longhand mathematics required to turn both their solar and stellar observations into positions.

With his observatory up and running, it took little time for Carrington to be recognized as one of Britain's greatest observers. Both John Herschel and George Airy held Carrington's observing skills in high regard. At one point, Airy requested his help in tracking down a mysterious error that had crept into the Greenwich data. It was, said the Astronomer Royal, as if the whole hill at Greenwich shifted, thus changing the positions of the stars. Carrington recognized immediately that the inconsistencies were due to the cooling down of the Greenwich telescopes after the heat of the day. He had already confronted this problem with his own telescopes and learned to correct it mathematically.

By 1855, it was obvious that Schwabe's predicted sunspot minimum was happening. John Herschel took this as his cue to lobby Carrington to convert his solar observations to the fledgling art of photography. He argued that a simple photograph could replace hours at the telescope sketching and timing. However, Carrington's sense of urgency over the importance of a consistent catalog of sunspots made him unwilling to change his methodology now that he had started. He estimated it would take three years to develop a solar photography telescope and perfect the technique to the accuracy with which he could draw the spots. Who knew what discoveries he might make during those years?

So Herschel transferred his photographic ambitions to Sabine's observatory at Kew. He must have found it comparatively easy to persuade the general to develop a telescope for making daily photographic records of the sunspots because such a device would allow Sabine to monitor his correlation between sunspots and magnetic storms directly, without relying on the data of others. Sabine jumped at the chance and approached wealthy businessman Warren de la Rue to oversee the development of the so-called photoheliograph. De la Rue's family business was in printing. He himself was already an accomplished amateur astronomer and had taken a great interest in the early photography of the Moon. Adapting the process to capture the Sun's image appealed to his sense of innovation, and he began the task.

Carrington continued sketching. The hours he spent studying the rolling solar surface made him more familiar with its characteristics and idiosyncrasies. In particular, drawing the intricate details of the sunspots convinced him that both William and John Herschel, for all their stature, were wrong. Sunspots were not gaps in the solar surface. The more closely one looked, the more detailed the darkest areas became. There was no way they could simply be wells into the interior of the Sun; they contained too much individual structure. Somehow they were dark features that floated in the solar atmosphere.

By 1856, the grueling nature of the day and night observations began to exact a price on both Carrington and Simmonds. Although Simmonds worked every clear night at first, regardless of the date, he eventually requested to be allowed Sundays off. Carrington agreed but continued to observe by himself. Eventually, even Carrington needed a break and awarded himself a holiday. He took a fortnight to visit Germany and its environs, following an itinerary that swept him past as many observatories as possible. In particular, he was determined to visit that of Heinrich Schwabe.

Carrington arrived in Dessau and found the observatory located in its center. At first sight, this appeared to be the most unsuitable place for an astronomical establishment. More curious still was that the address on St. Johannis Strasse was that of an ordinary house. Schwabe, a round-faced, balding man of sixty-seven, greeted Carrington and led him to the top floor. In a small attic, barely twelve feet by ten feet, whose windows looked out over the roofs of the surrounding houses, stood the most magnificent Fraunhofer telescope. It was pointed toward one of the southern-facing windows, where the sunlight streamed in around midday. This humble setting was the site of the profound solar discovery that had so intrigued Carrington.

For the last thirty-one years, Schwabe had climbed daily into the attic and logged the number of sunspots. On average the weather allowed him to observe the Sun for three hundred days out of every year. He showed Carrington his observing notes: 9,000 or so observations of around 4,700 sunspots, written into a library of bound notebooks. As well as facts and figures, Schwabe had also sketched the more interesting spots he had seen. He said that his discovery of the solar cycle made him comparable to Saul, who went in search of his father's asses, and found a kingdom. One thing eluded him, however. He confessed that he had never been able to accurately deduce the rotation rate of the Sun.

As the two astronomers talked about their respective solar observations, Carrington learned that Schwabe too had reservations about thinking of sunspots as holes. When the spots drew close to the edges of the Sun, it was well known that they appeared as depressions; this had been the observation William Herschel had proclaimed in 1795, while forgetting to mention that Alexander Wilson had announced it a quarter of a century before. Schwabe and Carrington talked about how their own observations revealed that the depth of the depressions varied from spot to spot. This was significant because if the spots were truly gaps down onto the dark surface of the Sun, they should all be the same depth.

Carrington left Schwabe, inspired by the single-minded nature of his devotion and the success he had achieved. Upon returning to England, Carrington recounted the expedition at a gathering of the Royal Astronomical Society and began to press for Schwabe to be officially recognized by the organization for the groundbreaking work. The following year, Schwabe was named the 1857 winner of the RAS's highest honor, the Gold Medal.

While announcing the award, the president of the RAS at that time, Manuel Johnson of the Radcliffe Observatory, University of Oxford, praised Schwabe for his indomitable zeal and untiring effort, stating that "the energy of one man has revealed a phenomenon that had eluded even the suspicion of astronomers for 200 years." That same year, Carrington became the RAS's secretary alongside Warren de la

Rue, who continued to experiment with solar photography at Kew. As befitted his new role, Carrington visited Schwabe again to personally present the medal, which was stamped with the likeness of William Herschel's forty-foot telescope.[1]

The year was also notable for Carrington because he finally completed the northern star catalog over which he had labored for four years. Although the observations had been complete in 1856, it was the mathematical reductions that had extended the project another year. Carrington estimated that Simmonds performed three-fifths of these and was clearly grateful for his efforts, referring to him as "my friend George Harvey Simmonds" in the catalog's introduction. Nevertheless, Simmonds left Carrington's employ shortly afterward.

The catalog itself was eventually deemed so important to the art of navigation that the admiralty published the volume using public money. The work established Carrington's name in astronomical circles. No longer was he just a promising newcomer, now he had actually made an achievement.

He did not rest on his laurels; instead he began the search for a new assistant and returned with fresh enthusiasm to his solar observations. There was a long-standing mystery about the speed of the Sun's rotation to be solved. During his trawl of previous sunspot observations, he had found that measurements of the Sun's rotation period varied from twenty-five to twenty-eight days. He was confident that his Redhill telescopes had already provided the data to pin down the time much more accurately.

He reasoned that as sunspots appeared in both solar hemispheres and were swept around by its rotation, they would travel parallel to the Sun's equator. The only problem was whether the spots themselves changed their position, as well as being carried by the rotation of the Sun. To get around this possible problem, he needed a large number of solar observations, so that any individual motion of the spots would be averaged away. He chose to discard any large or straggling group of sunspots because they changed their appearance on a daily basis, making it difficult to precisely determine a central point. Choosing only the

[1] Even today, William Herschel's forty-foot telescope remains the official logo of the Royal Astronomical Society.

roundest, best-defined, and most isolated sunspots, Carrington set to work collating the results.

As usual, it took months to perform the mathematics, but, in 1858, he realized why earlier determinations had been so diverse. It had nothing to do with the accuracy of the telescopes, but rather that spots at higher latitudes consistently moved more slowly than those near the equator. So earlier observers had probably been comparing the rotation of spots from different latitudes.

Carrington announced the results to the Royal Astronomical Society, explaining that the Sun did not rotate as a solid ball, but that the equator rotated once in about twenty-five days, whereas the midlatitudes took three days longer. The average rotation rate of the Sun was about twenty-seven days. This "differential rotation" was strong proof that the Sun was a wholly gaseous body because no solid object could rotate at different speeds. Yet when he presented his finding to the RAS, Carrington made little comment about the challenge his observations posed to the Herschellian idea of a solid Sun. He preferred to let his observations speak for themselves, so that others could draw the necessary conclusions. In fact, it was becoming obvious that he shied away from theory as much as possible, preferring to concentrate his efforts on observing, the area where he knew his considerable talents lay.

That day at the RAS, there was something else he did not mention, a discovery so full of potential that he did not want to say anything until he had double-checked his conclusions. He had noticed that the sunspots did not appear at random latitudes. At the beginning of a sunspot cycle, marked by the minimum number of spots, those that did appear stayed at higher latitudes. As the cycle continued and more spots appeared, they emerged at lower latitudes, edging closer to the equator. The behavior was mirrored in both hemispheres. It was such a potentially important aspect of Schwabe's solar cycle that Carrington decided to postpone announcing it until he had fully analyzed his data. He never really got the chance.

In July, Carrington's father died suddenly at the age of sixty-two. Carrington wrote immediately to the Greenwich and Durham observatories, imploring them to continue his sunspot observations while he made arrangements for his father's funeral. On arriving at the family brewery in Brentford, Carrington learned that there was no one but him to take over the business. His mother and mentally unstable younger brother, David, needed providing for. So Carrington was thrust headlong into the business of running a brewery and found little time to return to astronomy.

By November, he had still not completed the work to definitively prove the drift in latitude of newly appearing sunspots. Nevertheless, he could delay the announcement no longer. Warren de la Rue had perfected the photoheliograph and was taking daily photographs from Kew. Across Britain and Europe, other astronomers were also eagerly studying the Sun. Although Carrington had a substantial lead, his new duties would mean that the others would soon catch up with him. At that month's meeting of the RAS, he apologized to the Fellows for the sketchy nature of his proof before presenting the briefest summary of his suspicions. His fear of competition was well founded.

A few years later, the German astronomer Gustav Spörer announced the same conclusion with a rich analysis of independently collected data to back up the assertion. The drift, from high to low latitudes as the sunspot cycle continued, swiftly became an accepted aspect of the solar cycle, known as Spörer's law. Even though Carrington had announced it first, the German's fuller proof ensured it was Spörer's name that was connected with the phenomenon.

The loss of credit must have brought into harsh focus to Carrington that, unless he could juggle his life as a scientist with his responsibilities as a brewer, his career as an astronomer was over.

SIX

The Perfect Solar Storm, 1859

Adjusting to the new status quo was a slow process for Carrington. His widowed mother came to live at Redhill, and the astronomer commuted back and forth into London to the brewery. Fettered by the business, he did his best to maintain the strictest of observing schedules. He also continued in his role as joint honorary secretary of the RAS with Warren de la Rue, a position that afforded him a high profile within the astronomical community but leached away still more of his time.

Soon, he was forced to deputize some of the observations to an assistant. Entrusting his observatory to another might not have been so worrisome had he still possessed the support of the reliable Mr. Simmonds; but since that gentleman's departure, Carrington had struggled to find a worthy replacement. None of the subsequent assistants attained the standards of Mr. Simmonds. As a result, Carrington shouldered most of the time-consuming mathematics, as well as all the observations he could possibly manage. He became increasingly despondent that his research was falling irretrievably behind schedule.

In March 1859, he heard that the Oxford astronomer Manuel Johnson had died suddenly at the age of fifty-four, leaving vacant a well-paid position at the university's Radcliffe Observatory. At the news, Carrington decided to apply for the job. Once restored to full-time astronomy, with access to a staff of mathematicians to boot, he could sell the brewery to the highest bidder. The problem was that Carrington would have to leave his beloved Redhill and make himself answerable to the observatory's trustees—a situation that had ended disastrously for him at Durham.

He wrote to John Herschel, sharing his thoughts and appealing for the sixty-seven-year-old scientist to vouch for him. "If I am to leave Redhill I shall feel like an Arab parting with his favorite mare, I hardly know if we can part," he wrote. But he could see no other solution. He asked for Herschel's confidence in this matter as he realized just how strange it might seem for a gentleman to be giving up his independence to place himself in employment. "It might be misunderstood in this City where few believe that a man may prefer astronomy with a small income to a business with a larger one," he wrote. Until the job was his, he wanted no talk to be made of his desire, so that his reputation and judgment might remain unquestioned.

Of all the people to ask, Herschel was probably the most sympathetic to Carrington's predicament. He was only a few years into retirement from an unhappy career move in which he had left science to become the Treasury's Master of the Mint. The difficulties associated with the job had so thoroughly curtailed his ability to perform science that Herschel had retired in 1855 to once more resume his investigations of the natural world.

With Herschel's support, Carrington applied for the job and waited. Months passed without Carrington hearing a thing. The post lay empty and the work of the Oxford Observatory floundered. In the meantime, he struggled on at Redhill, little knowing that 1859 would become his most famous year.

First, he learned that his perseverance was to be rewarded this year with the Royal Astronomical Society's Gold Medal. Ostensibly this was for the epic catalog of 3,735 stars that he had compiled in just three years. Then, in late summer, he saw the spectacular solar flare when cataloging that day's sunspots, and that night the Earth cowered beneath the aurora as the magnetic effects of this perfect solar storm swept past Earth.

Already one of the world's solar experts, Carrington was crowned the Sun King for this extraordinary discovery, and he was elected to Fellowship of the Royal Society soon after with a list of proposers that reads like a "Who's Who" of Victorian astronomy. There was George Airy, the Astronomer Royal; John Herschel; the 3rd Earl of Rosse, who was building a monstrous telescope to rival the memory of William Herschel's forty-foot; John Couch Adams, the unassuming Cambridge theoretician whose prediction of Neptune had been ignored; James Challis, who had inspired Carrington to take up astronomy; and old friend Warren De la Rue, among fifteen others including Baden Powell, the founder of the scouting movement.

It should have been the high point of Carrington's career, but all it did was underline that he should have been devoting all of his time to astronomy. Unlike some of his peers who dabbled in astronomy as a relaxation, Carrington's passion was too strong to be satisfied by anything other than total immersion, and he struggled to find a way to balance himself between commerce and science.

The only advantage held by the brewery was that it rested just a few miles from the Kew Observatory where De la Rue was making huge strides in solar photography. While attending to business one day in early September, Carrington slipped across the Thames, past the Kew botanical gardens where plant specimens from the farthest shores of the empire were cultivated in cathedrals of glass, and up the long driveway to the brilliant white observatory building.

For Balfour Stewart, the recently appointed director of the Kew Observatory, Richard Carrington's visit with his description of the solar flare was a defining moment. He thought the magnetic blip that coincided with Carrington's flare was a clear signal that the Sun was affecting the Earth through magnetism. When he later showed the tracing to the Royal Society, he announced that, in recording the signal, "our luminary was taken in the act." However, understanding the mechanism behind the connection was far from easy.

Great magnetic storms had occurred before and after Carrington's flare. The first commenced in the late evening of 28 August and the other in the hours before dawn on 2 September, as reckoned by Greenwich Mean Time. The tracings of these two events dwarfed the blip and represented two almighty periods of magnetic chaos.

Stewart was convinced that all three events were somehow related and began collecting data from the other magnetic observatories around the world. All had readings of the two great storms, but no one else had seen the tiny spike that coincided with the flare. This was because most stations relied upon observers manually taking readings every hour or so. Indeed, had it not been for the continual nature of the photographic recording device, the Kew team themselves would also have missed this little spike of magnetic coincidence. Stewart thought it held the most breathtaking scientific potential. In rounding out a paper on the subject to the Royal Society, Stewart wrote, "if it be true that the spots on the surface of our luminary (or action connected with these spots) are the primary cause of magnetic disturbance, it is to be hoped, since the study of the Sun's disc is at present a favorite subject with observers, that ere long something more definite may be known with regard to the exact relation that subsides between these two great phenomena."

Stewart was not the only one fascinated by these magnetic storms.

In America, Elias Loomis, the professor of mathematics and natural philosophy at New York University, was preparing to move to Yale to pursue research into the new science of meteorology, but the unprecedented scale of the aurora of 28 August arrested his attention.[1] Having witnessed the display himself, he immediately recognized the importance of the event. Writing in the *American Journal of Science and Arts*, Loomis appealed for any other observations of the aurora or the magnetic storm, and of its effects on the telegraph lines. The journal was inundated. Loomis, who had isolated himself in work since the death of his wife Julia five years earlier, picked through eyewitness reports and built up a staggering picture of the global mayhem surrounding Carrington's flare.

It began on 28 August at around 6:30 P.M. when all telegraph lines out of Boston's State Street office stopped working. In other offices, the onset of the magnetic storm was much more ominous. In Springfield,

[1] One of Loomis's early attempts at meteorology was to estimate tornado wind speeds in a rather macabre way. A persistent prairie story was that chickens unlucky enough to be caught in tornados were often stripped of their feathers. In 1842, Loomis chose several even unluckier chickens for his experiment. He killed them and fired each carcass from a cannon. His plan was to use different explosive charges so that each chicken would be propelled at a different velocity, then to examine each to see which had been stripped and which had remained feathered. Things did not go quite according to plan. He wrote, "My conclusions are that a chicken forced through the air with this velocity is torn entirely to pieces; so tornadoes likely possess wind speeds of less than the measured chicken speed of 341 miles per hour." Say what you like about his methods, but you can't fault his reasoning.

Massachusetts, a giant spark crackled from the telegraphy equipment into a nearby metal frame, heralding the disintegration of communications for the night. The electrical arc persisted for so long that it filled the office with the smell of scorched wood and paint.

In Pittsburgh, Pennsylvania, operators struggled to disconnect the batteries from the lines as the auroral currents threatened to destroy the equipment. In the process of disconnection, not just sparks but "streams of fire" leaped around the apparatus, subjecting the delicate platinum contacts to the danger of melting. With their nimble disconnections, the operators saved the contacts but found the equipment too hot to touch. Not so lucky was telegraph operator Frederick W. Royce in Washington, D.C., who was stunned by a large arc of electricity that leaped upward and struck his forehead. He recovered soon after, but the event showed the potential for mortal danger.

All night, the operators struggled to send their messages. The best they could hope for were gaps of 30–90 seconds before the overwhelming waves of phantom electricity again took hold of their equipment. Outside these pockets of normal operation, the current on the lines either dropped to nothing or surged so powerfully that the armatures used to tap out the messages were gripped tightly by magnetism and could not be moved. When the lines did clear, little business was conducted because the operators were too busy gossiping with their long-distance colleagues about the unprecedented conditions of the night—and the superintendents were too busy documenting the extraordinary behavior of the equipment to check how much real work was leaving the office.

Everyone on the staff noticed that the longest lines carried the largest interfering currents, but on this night even the shorter lines were affected. The alien current noticeably affected the line that ran from Boston city center up to the Harvard Observatory just three miles away. Telegraph operators knew that electrical interference of this kind was accompanied by auroras, and many must have wondered whether the celestial lights would be as extravagant as the interference. They were not disappointed.

The telegraph had been disrupted shortly after 6 P.M., and, although twilight had not yet begun, the rosy hue of the aurora immediately suffused the sky, becoming more noticeable as the sky darkened. In Newbury, Massachusetts (42 degrees 48 minutes latitude), the aurora in the

An engraving of the aurora used to illustrate an article by Elias Loomis in *Harper's New Monthly* magazine, 1869. (Image: Stuart Clark, private collection)

east easily outshone the sunset in the west, and in Marquette, Michigan (46 degrees 32 minutes), the aurora formed floods of white light on the horizon that passed into "crimson fleecy vapors" at the zenith. Cocks crowed in Grafton, Canada (44 degrees 3 minutes), fooled by the aurora into thinking it was dawn. In Green Bay, Wisconsin (44 degrees 30 minutes), a gentleman by the name of D. Underwood described the full light show:

> The aurora was visible in the northern part of the heavens, but did not attract particular notice until about 9 P.M. Soon after

> eight the sky began to redden, and became nearly of a blood-red color. Soon the streaks were observed shooting upwards from all points of the horizon, and concentrating in a large luminous mass in mid-heavens. The greatest intensity of color was at the zenith. Rays were constantly shooting up from all points of the horizon and the colors constantly changing. The rays emitted an intense red light for about half an hour, when they began slowly to fade away in the north and south, but in the east and west they continued to glow until 10 P.M., when they began to fade away. Flashes of white light appeared amoung them, commensing from the horizon and moving upwards, following each other in rapid succession like the waves of an immense sea of light. They grew brighter as the red color disappeared, and when this was wholly gone they also gradually faded away.

This same display could be seen from Key West, Florida (24 degrees 33 minutes), as a blanket of fiery red light across the northern sky. In Inagua, Bahamas (21 degrees 18 minutes), a similar red sky glow sparked panic that a large fire was taking place somewhere in the neighborhood.

Although the lights faded from the sky before midnight, that was not the end of the aurora that night. The disruption to the telegraph continued, and in the early hours of 29 August, the sky was again transformed into a tremulous bowl of illumination. Professor C. G. Forshey had turned in for the night, believing that the fading of the first round signaled the end of the show. He chanced to awake at 3 A.M. and "perceived that it was very light outside, rose, and found the whole northern heavens again on fire."

William Dawson of Henry County, Indiana (40 degrees), witnessed the onset of this second bout:

> About midnight, a dark cloud decked with immense streamers of white, glaring light, rested on the northern horizon, when suddenly it burst forth into streaming coruscations of red, purple and white lights, shooting to a point 15° or 20° south of the zenith, where these flashing lights presented the appearance of a cloud, tinted with vermillion [*sic*] and purple. At 12 and a half A.M. fully two-thirds of the heavens were wrapt in flashing torrents of streamers.

This new flood of color was described by many as giving the same light as the Moon and allowed them to read the larger print found in newspapers. Back in Newbury, Massachusetts, this second bout was so bright that the stars were lost amid the aurora. In Sacramento, California (38 degrees 34 seconds), "The whole northern sky . . . seemed to be a cupola on fire, supported by columns of diverse colors, relieved and intensified by dark shadows."

In Britain, night had fully spread its darkness over the land when the assault began at 10:30 P.M. on 28 August. As the Kew magnets jumped, a burning purple arch appeared across the sky, attended by powerful streamers and curtains of red and orange light. Across Europe the story was the same, with vivid auroras and the breakdown of telegraphic communications. Only Athens in Greece (38 degrees 2 minutes) escaped with no reported sightings despite clear weather. In many places the auroras were only banished from the sky by the daylight. The disruption to the telegraph, however, continued all day, testifying that Earth's atmosphere was still ringing with electrical and magnetic energy.

As the Arctic lights had smothered the northern hemisphere, so their southern counterparts reached up from the antipodes. In Australia, the aurora was recorded at the Sydney Observatory (33 degrees 52 minutes south) as a bright red glow with streamers in the southern sky. To his chagrin, one astronomer missed the display entirely. Richard Schumacher, assistant to the Chile Observatory, slept soundly in his berth on a ship near Cape Horn (57 degrees south) during the display of 28 August. On hearing the sailors talking the next morning, he begged the mate to awaken him should the aurora return. On 2 September, Schumacher was roused from slumber in the early hours of the morning. It was happening again.

Scientists at Kew, and the other magnetic observatories around the world, had been watching the uneasy rhythm of the Earth's magnetic field since the 28–29 August display. They knew that whatever was happening, it was not over. During the night of 1–2 September, in the aftermath of Carrington's flare, the auroras exploded into common view again and this time they were bigger and more sustained than even the 28–29 August display.

In an encampment at Sierra Abajo, Utah (37 degrees), Dr. John S.

Newberry was awoken by a red light that penetrated his tent. Outside he saw the sky wreathed in bright crimson with streaks of white and yellow light converging on the zenith. In Cahawba, Alabama (32 degrees 25 minutes), the colors waved like a gigantic pennant in the wind. In St. Petersburg, Russia (latitude 59 degrees, 56 minutes), the magnetic variations were so abnormal that the usual hourly readings were replaced with ones taken every five minutes. Between 1–3 September, the Russian magnetic instruments were overwhelmed by the strength of the magnetic storm, and no useful measurements could be taken. The story was the same at the other magnetic observatories around the world;[2] surges of magnetism during the storm pushed the readings off the scale, with the result that the maximum strength of the storm could only be estimated.

For the telegraph operators, another day of disruption and danger beckoned. Some decided to fight back. George B. Prescott was the superintendent of Boston's State Street telegraph office. He had kept careful notes about the effects of the aurora on the telegraph lines for almost a decade, having first heard stories back in 1847 and then witnessing some feeble effects in 1850. In 1851, he witnessed his first powerful display when the aurora took full possession of the lines. A year later he witnessed the danger. The office where he was working used the Bain electrochemical method of recording incoming messages. This consisted of a set of papers, prepared by having been soaked in a solution of potassium, nitric acid, and ammonia. This rendered the paper susceptible to electricity; positive polarity decomposed the chemicals on the paper and stained a spot blue while negative polarity bleached the paper. Toward the evening of 19 February 1852, Prescott watched as a blue line appeared on the paper, indicating that a constant current was passing through the wire. The line grew darker as the current increased, until the paper caught fire. Fighting against the chemical-laden smoke, Prescott put out the fire and watched the current die away. Instead of stopping at zero, the current proceeded to grow in negative polarity until it again sparked a fire.

[2] Although British government funding for the observatories in General Sabine's magnetic crusade had elapsed in 1849, many others were still operating on private and provincial funding.

As he observed the auroral effects on 28–29 August 1859, Prescott noticed that, in between the surges, the auroral current was often of comparable strength to that supplied by the batteries. He conceived a plan. A telegraph line worked, using two batteries, one at either end. The batteries were connected to the ground and to the wire via a finger-operated armature that could make or break the circuit. By tapping the armature up and down, electricity pulsed down the wire. When the aurora was overhead, electricity was constantly on the wire. He wondered, why not disconnect the batteries and just work with the phantom current? He hastened his ideas into print in the 31 August edition of the *Boston Journal* but did not think he would have a chance to test them so soon.

As business commenced on 2 September, the lines were almost unworkable thanks to the magnetic storm. At Prescott's suggestion, the Boston operator asked the Portland operator to disconnect the battery and connect the telegraph line through the armature, directly to ground. After doing the same, the Boston operator sent, "We are working with the current from the Aurora Borealis alone. How do you receive my writing?"

"Very well indeed, much better than with the batteries on. There is much less variation in the current and the magnets work steadier. Suppose we continue to work so until the Aurora subsides?" asked the Portland operator.

"Agreed," replied the Boston operator and commenced sending the day's dispatches. Other operators across the world independently reached the same conclusion, and they too soldiered on through the day.

As he pored over the plethora of reports, Loomis saw that, like magnetic storms, auroras were global events. They occurred everywhere they were going to be seen, virtually at the same time. He charted the positions of the aurora on a map and realized that the northern light show had occurred across a broad oval expanse, offset from the Earth's rotation axis but tilted in such a way that it encircled the northern magnetic pole, in the Canadian Arctic. This offset tilted the auroral oval down across North America into the western hemisphere's tropical zone but lifted it away from the eastern tropics. Although the reports from the southern hemisphere were sketchier, there was little doubt that a vibrant display of similar size had occurred there too, although most of it fell beyond the sight of man, taking place over the heaving waters around Antarctica.

Loomis triangulated reports of colored arches and streamers seen from widely separated locations to calculate the height of the aurora. He estimated that the base of the aurora, characterized by the arch, lay some fifty miles aloft. The streamers rose from the arch, like defensive spikes from the spine of a dinosaur, and soared upward to five hundred miles in height. In width, they varied between five and twenty miles. Everywhere the streamers were reported, they pointed roughly north–south.

His fascination with auroras soon extended beyond these most recent displays and he began searching for previous sightings. He swiftly saw that northern auroras were always attended by southern ones, and both occurred in belts that encircled the polar regions. The northern belt lay across Hudson Bay and a great swath of Canada. It continued through Alaska and across the Bering Strait to crown the Russian Empire before returning above the Atlantic, engulfing Iceland and the tip of Greenland. From within this belt, auroras could be seen on almost any clear night. The farther from this belt, the less likely you were to see an aurora. From as far away as Havana (23 degrees), Loomis could find just six auroral sightings from previous centuries. South of Havana, the number was negligible. On the other hand, north of Havana the auroras rose in both frequency and brilliance and became increasingly likely to engulf the whole sky. So Loomis concluded that auroral frequency rose with latitude. The farther north, the more frequent were the auroras. On the shores of the Great Lakes, you could expect to see a few dozen per year.

All his work reinforced the unprecedented scale of the 1859 auroras, and the tales of telegraphic danger highlighted the frightening nature of the event. Man's dominion suddenly felt less assured. The Earth had been caught in something inexplicable, the first phenomenon seen that had a direct influence on Earth but had nothing to do with gravity—the only force that had been thought capable of communicating itself across space. At the heart of it all was Carrington's flare. Suddenly, it became imperative to uncover the flare's ability to cause auroras and magnetic storms.

For his part, Loomis firmly linked the auroras to the telegraphic disruption. A number of the observers had described the movement of auroral features or stated that the lights shimmered like a flag in the wind. From their accounts, Loomis found that the movement swept from northeast to southwest. This was the same direction as the waves of auroral current that swept across the telegraph lines, and he concluded that the aurora was produced by the flow of electricity through the atmosphere. Balfour Stewart also reached substantially similar conclusions, believing that the magnetic storm was caused by a flowing current of electricity that swept past the entire Earth, expelled from the Sun by the solar flare.

Other scientists wondered what affect the auroras had on the weather. Most assumed there must be some connection because both appeared in the atmosphere. They wondered if there was a link to thunderstorms, as these were the other occasions when electricity filled the air.

As they struggled to find a link between atmospheric, solar, and magnetic phenomena, astronomers soon found themselves mired in cause and effect. Why did Carrington's flare only produce a small disturbance compared to the magnetic storm eighteen hours later? Was the subsequent storm linked to the flare somehow, or did an even larger flare that went unseen take place eighteen hours later? If every magnetic storm required a flare, why weren't more flares being observed? This was a big puzzle because the number of solar observers was growing all the time; surely someone would be looking in the right place at the right time.

In hunting for even the smallest clues to understand this unanticipated facet of cosmic interplay, astronomers began to reinvent themselves. The passive measurement of celestial positions and movement that had hitherto defined their science would not help in this new quest. They needed to reach across space and divine the very nature of the Sun: what it was made from, what reactions took place inside it, what made it shine, and, of course, what triggered the solar flares.

But how could they do this?

By coincidence, in the very same year, scientists in Germany solved an entirely separate solar puzzle, one that had been vexing astronomical minds for the last fifty-eight years. In doing so they gave astronomers a new and powerful window onto the cosmos. If the effects of Carrington's flare gave astronomers the motive to investigate the nature of the Sun, the work of Gustav Kirchhoff and Robert Bunsen, in Heidelberg, gave them the means by which to carry it out.

SEVEN

In the Grip of the Sun, 1801–1859

The mystery had begun back in 1801, by coincidence around the same time that William Herschel had been trying to engage the scientific community in a discussion about the nature of the Sun. English chemist William Wollaston passed sunlight through a prism and projected the resulting spectrum onto a wall some ten feet away. He saw that, cutting across the colored rainbow of light, were four vertical dark lines. Wollaston assumed that these lines simply represented natural gaps between the colors, and there the matter rested until twenty-seven-year-old Joseph von Fraunhofer rediscovered the lines in 1814.

Fraunhofer was a man with a single purpose in life: to produce the finest glass in the world. He had been orphaned at eleven and forced into the servitude of glasscutter and mirror maker Philipp Anton Weichselberger. Three years later, he was buried alive when Weichselberger's workshop collapsed. Somewhat amazingly, this proved to be a fortuitous event because the Prince Elector of Bavaria, Maximilian IV Joseph, was present when the teenager was pulled unharmed from the wreckage after four hours of digging that had already revealed the crushed body of Weichselberger's wife. Something about Fraunhofer captivated the prince, who supplied him with books and insisted he be allowed time off to study. Also present at the rescue was Joseph Utzschneider, a politician and entrepreneur. He too encouraged Fraunhofer's ambitions.

Under their combined patronage, Fraunhofer studied hard and within eight months had won a position at Utzschneider's Optical Institute at Benediktbeuren, a former monastery that now specialized in glass making. Leaving Weichselberger's regime of virtual slavery behind,

Fraunhofer pursued the creation of glass at Benediktbeuren with the passion of an alchemist, mixing molten metals into the liquid glass to produce telescope lenses that became the envy of the world.

While testing his lenses to see how they dispersed the natural colors, Fraunhofer rediscovered the dark lines within the solar spectrum. As he studied them, he found that the pattern never changed and that some lines were deep and black, but others were merely faded bars of color. Fraunhofer labeled the eight darkest A to H. In between the prominent lines B and H, he counted 574 other lines of varying strength and carefully set about recording their positions.

Throughout his career he returned to the solar lines. In 1823, he used a newly invented device of his own making, known as a diffraction grating, to produce a more precise spectrum than a prism. This allowed him to see the lines as never before and to accurately measure the wavelength of the darkest lines.

Next, Fraunhofer turned his diffraction grating to the brightest stars, revealing that they too showed dark lines in their spectra. The positions of the lines shared both similarities and differences to each other and to the Sun. What were these lines? What did they mean? Some believed they were defects in the telescopes, others that they were produced in the atmosphere of the Earth or, more tantalizingly, in the atmospheres of the Sun and other stars. Before Fraunhofer could reach an answer, tuberculosis ended his life. He was just thirty-seven.

Various scientists then chipped away at the puzzle of the Fraunhofer lines, as they became known. In the year of Fraunhofer's death, John Herschel and his collaborator, William Fox Talbot, realized that each chemical element gave off a unique pattern of colored lines when burned. They wrote, "A glance at the prismatic spectrum of a flame may show it to contain substances which it would otherwise require a laborious chemical analysis to detect."

As knowledge of such "flame tests" spread, scientists speculated that perhaps the dark Fraunhofer lines were somehow related to the bright lines produced when chemicals were burned in the laboratory. If so, the Fraunhofer lines could reveal the presence of chemical vapors in the atmosphere of the Sun and other stars. If astronomers could determine which lines were produced by which vapors, they would gain a power beyond belief: the ability to deduce the chemical composition of celestial objects.

The influential French philosopher Auguste Comte thought the endeavor sheer folly. In 1835 he wrote, "We understand the possibility of determining their shapes, their distances, their sizes and their movements; whereas we would never know how to study by any means their chemical composition." Others thought his pessimism profoundly misplaced and drove on with their investigations.

In the race to understand the chemistry of space, John Herschel and others managed to photograph the solar spectrum showing the Fraunhofer lines during the 1840s. Herschel also showed that the lines extended into the infrared region of the spectrum by wetting a strip of paper with alcohol and laying it in the invisible spectral band that his father had discovered beyond the visible colors. John watched it dry in stripes, which he concluded were caused by the infrared equivalent of Fraunhofer lines.

Meanwhile, Fox Talbot studied lithium and strontium, both of which burned with a red flame. On passing this light through a prism, he discovered that both resolved into different patterns of red lines. So even when the color of the flame could not distinguish between chemicals, the spectrum could. Others realized that Fraunhofer's D line was linked to the element sodium and that there was a suspicious similarity of potassium's red lines to a group of dark lines clustered around Fraunhofer's A line. In spite of this progress, there were problems.

Pure chemicals were difficult to manufacture, and the contaminating elements gave off their own lines, ruining the unique patterns that would otherwise be seen. Sodium, in the form of salt, was a particular terror because it contaminated everything, and the merest trace was enough to produce a vibrant yellow line at Fraunhofer's D position. There was also a show stopper: why were the Fraunhofer lines dark, but the flame test lines bright? Until this could be explained, the analysis would feel too much like magic for anyone to be confident about applying it.

It fell to physicist Gustav Kirchhoff and chemist Robert Bunsen to realize the connection between the dark and bright lines. Bunsen and his laboratory assistant Peter Desaga had perfected a gas burner that nowadays carries the professor's name for flame testing, whereas Kirchhoff was a talented physicist. Bunsen tempted the younger Kirchhoff to join him at the University of Heidelberg where they could collaborate. Both recognized in the other complementary skills and set about uniting the solar spectrum and flame tests once and for all.

Bunsen used his chemistry to produce chemical samples of unprecedented purity, allowing their individual spectral lines to be gauged with confidence. Kirchhoff used his physics to design incomparable equipment with which to analyze the lines. Disabled from an accident earlier in life, Kirchhoff fussed over the delicate equipment in the darkened laboratory, managing to direct light rays into his apparatus with exquisite precision. His breakthrough came the day he burned a sample of lime, producing the famous limelight that illuminated theater stages. The substance burned with an incandescent flame that produced a continuous sweep of colors when split with a prism. Before hitting the prism, however, Kirchhoff focused a sliver of limelight through the flame of one of Bunsen's burners. He then peppered the flame with a sprinkling of sodium, making it flare with the chemical's characteristic yellow light. On the screen, he saw Fraunhofer's black D line appear in the spectrum of the limelight. The sodium vapor had absorbed that specific yellow wavelength from the limelight and blazed it around the lab in the form of a yellow flame.

He tried it next on sunlight, only this time he used a chemical that did not have a corresponding Fraunhofer line, lithium. Dusting the Bunsen flame with lithium powder, he watched in fascination as a dark lithium line appeared in the solar spectrum. At a stroke, Kirchhoff proved two things: there had to be sodium in the Sun because of the presence of Fraunhofer's D line, but lithium was missing because of the lack of a lithium line. He had done what Comte believed impossible: investigated the chemical composition of an object without actually having chipped off a piece for analysis.

Kirchhoff toiled on in an effort to generalize the concept of spectral analysis, so that others could use it with confidence. Eventually he was certain of three things. First, a hot solid object or a hot dense gas produces a continuous spectrum, a solid band of color that runs through the rainbow from blue to red. Second, a hot tenuous gas produces an emission spectrum, a sequence of brightly colored lines with positions that depend on the chemical composition of the gas. Third, a hot solid object surrounded by a cooler tenuous gas gives an absorption spectrum, an otherwise continuous spectrum from which certain wavelengths have been absorbed, creating a series of dark lines with positions identical to the gas's emission lines. These rules provided the certain link between emission lines and absorption lines.

As word of this breakthrough spread throughout the world, astronomers took their first tentative steps in the new technique of spectral analysis. They found the metals iron, calcium, magnesium, and many others present in the Sun.

Beyond the Sun's chemical composition, Kirchhoff's laws provided a couple of inescapable conclusions about the nature of the Sun. Laboratory experiments showed that metals needed high temperatures to melt and give off vapors. So the Sun's atmosphere had to be fiercely hot, at thousands of degrees to sustain its atmosphere of metallic vapors. The Sun itself had to be even hotter, in order to give off the continuous spectrum of colored light that the atmospheric gases then absorbed to create the Fraunhofer lines.

Astronomers called the visible layer of the Sun the photosphere and recognized definitively that it was not a luminous cloud layer surrounding a solid object. Nothing could remain solid at temperatures in excess of thousands of degrees.

Above the photosphere was the solar atmosphere whose metallic vapors produced the Fraunhofer lines as light from the photosphere escaped through it into space. The only confusing factor was that the sunlight then had to pass through the Earth's atmosphere. This meant some of the Fraunhofer lines had to be bogus, produced by chemicals in the Earth's atmosphere rather than the Sun's. As more astronomers studied the solar spectrum it became apparent that the Fraunhofer lines split into two camps: those that were unchanging, and those that varied slightly in their darkness during the day. Astronomers soon realized that the variable lines depended upon where the Sun was in the sky. When it was lower in the sky, its light passed through more blankets of the Earth's atmosphere and some of the Fraunhofer lines deepened. These lines therefore represented the chemicals in Earth's atmosphere. The lines that remained steadfast betrayed the chemicals in the Sun's mighty atmosphere.

Spectral analysis became the new astronomy among the wealthy amateurs. One man who should have been at the forefront of this revolution was Richard Carrington. His flair for technological invention and his careful observations should have cast him in the role of pioneer. Instead, he was in the throes of a personal crisis.

EIGHT

The Greatest Prize of All, 1860–1861

Beneath the veneer of Carrington's scientific success was the consuming rot of family duty. Following the death of his father, he had hoped to increasingly off-load the day-to-day running of the brewery to underlings and return to astronomy. As one year slipped into another, however, he realized that the business demanded his full attention. Shackled by these responsibilities and forced to commute between Redhill and Brentford, he struggled to maintain his solar observations. Every day he did successfully record the sunspots, the new data only increased his mathematical workload.

Carrington's situation did not get any better when he finally heard from the trustees of the University of Oxford's Radcliffe Observatory. It was now fifteen months since the premature death of Manuel Johnson and Carrington's first application for the job. Two others had applied as well. The first was Robert Main, the fifty-two-year-old first assistant at Greenwich, who had succeeded Johnson as president of the RAS and had awarded Carrington his Gold Medal for the Redhill catalog. The second was Norman Pogson, age thirty-one, who had previously worked as Johnson's assistant at the Radcliffe, but had left when the salary could no longer feed his growing family.

In a strange move, the trustees readvertised the prestigious position without calling any of the original applicants to interview. The trustees blamed the hiatus on seeking advice for the direction of future work at the observatory. They also took the opportunity to reduce the originally quoted salary of £600 to £500. When the new advertisement inspired no further applicants, they made their decision, and Carrington learned that his bid had been unsuccessful. Robert Main was awarded the job. Still none of the applicants had been called to interview; instead, the trustees had been guided by a three-page recommendation from George Airy, under whom Main had worked for the previous twenty-three years.

Airy wrote to the unsuccessful Pogson, offering a frank assessment of his favoritism in lieu of an apology, "Mr. Main's claims on me . . . are like those of a son on the head of the family. This almost prevents me from saying a word in favor of any other person."

To Carrington, Airy sent nothing. In truth, the relationship between them had become strained. Carrington's habit of commenting on the work of Greenwich, which Airy had welcomed at first, had somehow crossed the line into unwanted criticism. Stories were circulating that the two of them had crossed swords at council meetings of the Royal Astronomical Society. Underlining his enmity, Airy eventually resigned from the RAS Club, a coterie of Fellows who dined together before the RAS meeting proper, because Carrington "coolly lighted a cigar" at one of the dinners.

There may have been another reason that made it easier for the trustees to overlook Carrington. The Radcliffe Observatory was strapped for cash, and they were well aware that some of the equipment needed modernizing if they were going to keep up with Greenwich and Cambridge. Given Carrington's precipitous exit from Durham eight years earlier over his fight for newer equipment, the trustees might have felt him incapable of working within the meager budget they had available.

Whatever the reason, their rejection only deepened Carrington's state of crisis. For him, remaining an astronomer was now the greatest prize of all. Once salaried, he would dispose of the albatross of his father's brewery and rededicate himself to science, particularly his solar observations. Until that time came, the work of the other astronomers flowed on around him.

Warren De la Rue was now routinely taking pictures of the Sun at Kew, using the photoheliograph instrument that he had devised. This unique telescopic camera captured in a moment what Carrington had labored for hours to sketch and measure. De la Rue was so convinced of photography's place in astronomy that he dreamed of an exploit that would prove its value once and for all.

There was a total solar eclipse coming, visible from Spain. In the scant minutes of totality, when the pallid streamers of the Sun's outer atmosphere were revealed, observers always struggled to record the details quickly and accurately. Sketches and eyewitness accounts always differed. De la Rue wondered if the photoheliograph could capture the moment when the Moon blotted the light from the Sun and plunged the surroundings into darkness.

He approached Carrington, and others who had seen the 1851 solar eclipse from Sweden, asking for advice. In particular, he wanted to know how bright the white corona and the pink flames, or prominences as they were being called, would be. The answers did not encourage him. According to the astronomers he talked to, they estimated that the Sun's atmosphere would glow no more brightly than the full Moon. This was worrisome because photography was in its infancy and not yet highly sensitive to faint subjects, so De la Rue decided to test the photoheliograph. He waited for the next observable full Moon and exposed several plates. Developing them immediately, his ambitions took a blow. Not the slightest impression of Earth's neighbor could be detected on the photographs.

Even though his chances of success were slim, he decided to transport the equipment to Spain anyway. At least he should be able to capture images of the partially eclipsed Sun, even if the full glory of totality escaped his photographic clutches.

Although Spain was relatively close to Britain, travel to the Iberian peninsula was uncommon. As a result, commercial traveling both to and within Spain was rare and expensive. An added hindrance was the high customs charges for bringing equipment into the country. De la Rue expected to fund the expedition from the profits of his printing firm, but was approached one day by George Airy. The Astronomer Royal felt that the eclipse was sufficiently important to begin lobbying government for help. He asked De la Rue to put together a budget for transporting the Kew equipment. The Astronomer Royal then requested a ship from the admiralty to transport a number of astronomical teams and their equipment to Spain. He asked the British government to negotiate the waiving of customs charges for the expedition. In return, he personally vetted the astronomers wishing to make the trip, requiring each to submit a detailed plan of their viewing site, accommodation needs, and scientific agenda. Characteristically, Airy took a very dim view of any potential eclipse watcher whom he judged to be merely interested in observing the picturesque qualities of the Sun's temporary disappearance. If the government were footing the bill, he wanted tangible scientific results to prove their money had not been wasted.

At the next meeting of the Royal Astronomical Society, the gathered Fellows swapped information and tips about the upcoming eclipse. Although Carrington was staying behind, shackled by responsibility, he exhibited a specially designed eyepiece that would make measuring angles during the brief minutes of totality easier and therefore faster.

Based on his own experiences at the 1851 eclipse, he had already written a pamphlet to prepare observers for eclipse observations. In particular, he wanted observers to look out for any activity related to sunspots and made it clear that the reason for observing the Sun was to "arrive at the secret of the true cause of the prodigious radiative power of the Sun." Astronomers knew that the Sun was not a burning ball or a chemically powered inferno; such processes simply could not produce enough energy. The source of power remained a mystery beyond conception, and would do so for decades until the concept of nuclear power literally exploded into human imagination during the twentieth century.

In the pamphlet, written before Airy and Carrington fell out, Carrington discussed an unresolved eclipse controversy. In 1836, Francis Baily, one of the Royal Astronomical Society's founding members, was watching an eclipse. As the Moon completed its position over the Sun's disk, the final sliver of sunlight broke into jeweled beads. Following Baily's description, other eclipse observers began to report seeing such Baily's Beads. Airy, on the other hand, never saw the colored dots, and Carrington unwisely attributed this to Airy's superior astronomical skills, opining that Baily's Beads were the result of defective telescopes. He wrote that the flaw was so notorious that on many an instrument might be engraved, "Warranted to show Baily's Beads." In reality, Airy was a poor observer who delegated all such duties to his staff at Greenwich. Baily's Beads were real, their beauty caused by the final rays of the Sun shafting through the valleys between the lunar mountains.

Taking another page out of Airy's book, Carrington emphasized the need for objective scientific reporting, noting that the strange feelings of dread felt by some observers had been duly noted and so, "persons who may witness an eclipse will perhaps spare us the unnecessary dilation of their feelings or the state of their nerves, beyond what may be necessary for another to judge whether they were sufficiently masters of themselves for theirs accounts to be trusted." He then disclosed that he himself had had harbored none of the fearful feelings at the 1851 eclipse and wondered whether this was because he had been forewarned.

Beyond the solar observations, there was considerable excitement among the astronomical community that the eclipse might enable the discovery of an inner world, widely thought to exist closer to the Sun than Mercury. If so, it would be an historic sight to see because the planets Venus, Mercury, Jupiter, and Saturn also happened to be close to the Sun and would all pop into view at the moment of totality. In preparation, Carrington reviewed his collection of observations from astronomers around the world who claimed to have seen anomalous silhouettes against the surface of the Sun, any of which might have been the putative world.[1]

He also communicated a mathematical recipe for deducing the coordinates of sunspots as recorded in a three-volume set of solar observations that had been made by J. W. Pastorf, a German astronomer from Drossen, and that John Herschel had recently donated to the Society's archives. He beseeched the gathered Fellows that someone with "more leisure time" take up the challenge of performing the mathematics so that the collection of latent data, which spanned the years 1819 to 1833, could be transformed into workable knowledge. He admitted that there was no probability of his being able to devote the time needed for the task, as he was struggling to keep up with his own program of solar observations. This admission must have come hard to Carrington. When a large set of data had been uncovered that duplicated some of the stars he was measuring for his star catalog, he had made the time to blend it into his own work. Only the year before, in receiving the RAS's

[1] It would take until the twentieth century for astronomers to let go of their belief in the intramercurial world, Vulcan. Einstein's general theory of relativity finally explained Mercury's peculiar orbit, negating the need for a gravitational tug from an unseen planet.

Gold Medal for the catalog, he had been praised for this diligence. Now he had to beg his peers for help.

As for the coming eclipse, some scientists speculated that, given the magnetic mischief associated with the Sun following Carrington's flare, there could be some noticeable magnetic effect at the moment of the eclipse as the Moon blocked out the Sun's influence on the Earth.[2]

Having secured the pledge of government support, Airy informed De la Rue that the Royal Navy's HMS *Himalaya* would transport the various teams of astronomers and their apparatus to Spain. De la Rue set forth with his preparations. He selected a staff of four to accompany him on the trip and supervised the construction of a wooden observatory that could be easily dismantled for transport and reassembled in Spain. The small building housed the telescope and served as a makeshift darkroom in which to immediately develop the photographic plates. They fitted a sink and cistern to the developing room and draped a large canvas over the whole affair, giving it the appearance of a tent. Then they carefully numbered every component and "flat-packed" it for the journey. The only thing that they could not take was the cast-iron mount on which the photoheliograph sat. It was simply too heavy, so De la Rue had a new mount made, again of cast iron but in pieces that bolted together and could be carried separately.

On 5 July, De la Rue sent all two tons of apparatus on its way to Plymouth, where he joined it a day later. De la Rue, his team, and several dozen astronomers from other teams spent the night getting used to the berths on the *Himalaya* and then set sail for Spain on the morning tide of 7 July 1860. Two days later, a small steamer rendezvoused with them and guided the naval vessel into Bilbao Harbor.

While the astronomers rested that night, enjoying the hospitality of Bilbao before dispersing into the countryside, De la Rue's equipment began the seventy-mile trek from the shore to his chosen observing site. Initially he had chosen the old Roman settlement of Santander but was advised to journey to the southern side of the Pyrenees to avoid the sea mists that plagued the shorelines of the Bay of Biscay. So he

[2] No effects were detected, probably because the instruments were not sensitive enough. Modern instruments do register an effect caused by the Sun's ultraviolet radiation being blocked by the Moon because the ultraviolet would otherwise strike the Earth's ionosphere, promoting magnetic effects there.

plumped for the farming village of Rivabellosa. Only when his party began the journey the next day did he realize exactly what his choice of site entailed. The track was torturous. As the party bumped along, traveling overnight and into the next day, quite apart from his own personal discomfort, De la Rue thought frequently of the delicate equipment that was similarly bouncing along. In particular, he feared for the three chronometers. Although he had personally packed each in a wooden carriage box and stuffed it with protective wood shavings, he had not anticipated that the rigors of travel would be this bad.

Upon arrival in Rivabellosa, he opened the chronometer boxes and found that his worst fears appeared to have been realized. One had been severely disturbed by its passage. The glass plate over the clock face had been shaken off, although thankfully not shattered, and the chronometer itself had worked out of its housing. A careful inspection of the clock hands showed that they were undamaged, and De la Rue carefully reassembled the delicate timepiece.

Despite this setback, everything else proceeded smoothly. He saw exactly the location in which to erect his wooden observatory. It was a flat expanse of compacted Earth used for threshing the wheat. The party's translator inquired into the hire of the thrashing floor only to find that the harvest had just begun and that the space would be used for its intended purpose the next day. When the translator explained to the farmer the historic intention of the expedition, the farmer immediately agreed to do his threshing elsewhere and, what's more, refused remuneration for the use of his land.

The party had a week before the eclipse and set about their preparations. They were joined by a local servant boy called Juan, who proved a fast learner and helped in any way he could. Together they constructed the observatory on the thrashing floor and installed the photoheliograph. They erected a second telescope for De la Rue to look through. He would sketch the eclipse as insurance, being still doubtful that the photoheliograph would capture the faint traces of the Sun's atmosphere.

With everything built, their next job was to determine precisely the coordinates of their observing site. When the Sun was overhead they checked the clocks, which had been set to Greenwich Mean Time before they left, to see how close they read to noon. The discrepancy allowed them to calculate their longitude. At night, they measured the altitudes of the brightest stars to calculate their latitude. With their position firmly fixed, they began dry runs for the day, observing the Sun, taking and developing photographs. They discovered that the canvas of the tent had to be doused with water to keep the darkroom cool; otherwise the Spanish heat fogged the photographic plates, spoiling the images.

As news of the encampment spread, more and more people became inquisitive about the work and the spectacle of the coming eclipse. The local officials came to visit the astronomers and warned them to expect a large crowd of onlookers on the day. In anticipation of keeping the throng at bay, the officials promised to send guards.

With two days to go, the weather closed in. Violent thunderstorms shook Rivabellosa almost without interruption. De la Rue watched the display with a mixture of awe and dejection. The next day, too, the sky remained enswathed in cloud, giving just the briefest glimpse of the Sun's disk around midday.

The eclipse day of 18 July 1860 arrived and the men anxiously watched the gray covering. De la Rue could not rest for anxiety about the uncooperative weather. All the preparation, work, and expense looked as if it were coming to nothing. Then at 10 A.M. he saw the first small band of clear sky, and the predicted crowd began to gather around the makeshift observatory. At noon, two hours before the predicted start of the eclipse, the sky suddenly cleared. There was little wind to blow the clouds away; they simply dissolved, leaving behind a brilliant expanse of blue sky and the yellow disk of the Sun.

To express his thanks for Juan's help, De la Rue took a spare slip of glass and smoked it with a match. He presented it to the local boy, so that the servant could watch the eclipse through the charred glass. De la Rue then took his place at the telescope.

More and more onlookers arrived as the astronomers went about their final, nervous preparations. The chatter of the crowd grew to cacophonous levels, masking the tick-tock of the large box chronometer that De la Rue had been hoping to listen to in order to time the eclipse events. He decided to take glimpses of his pocket watch instead. He caught sight of Juan, lost in feverish activity, smoking other slips of glass for the clamoring bystanders. The boy held the matches under the glass as long as possible before flinging away the phosphor sticks, lest they burn his fingers.

Warren de la Rue, to the left with his back turned, and his team at Rivabellosa prepare for the coming eclipse. The front has been removed from the observatory, showing the Kew photoheliograph in place. The doorway into the darkroom can be glimpsed to the right. (Image: Royal Astronomical Society)

Shortly afterward the ground shook with hooves as five mounted guards rode into the village. They placed themselves at the disposal of the astronomers and defined a perimeter around the thrashing floor, forbidding the two hundred-strong crowd to pass. The guards were a welcome sight; before their arrival, curiosity had overwhelmed a few of the bystanders who crept to the wooden observatory to peer inside at the outlandish equipment and its human masters before being shooed away by the astronomers.

With twenty minutes to go, an alarming smell arrested De la Rue's attention. Something was burning. The crackling of a fire drew his gaze. Juan's discarded matches had set fire to some loose corn, lying on the thrashing floor. Grabbing the water bucket used to wet the tent, De la Rue made haste to extinguish the fire before it spread any closer to their vulnerable wooden observatory.

Returning to his position he waited anxiously. As the appointed time approached, he asked his staff to prepare the first plate. They set the chemicals and loaded them into the photoheliograph, but the moment of the Moon's first contact with the Sun did not arrive as expected. Puzzled, De la Rue checked the time on the clocks and revealed his mistake. The pocket watch was fast by some 8 minutes 11 seconds. With horror he realized that the chemically delicate plates would be ruined in that time and he ordered his team to begin preparing new ones as fast as possible.

He watched the Moon's limb take its first tentative bite out of the Sun at 1:56 P.M. but the plates were not ready until 2:02 P.M., when they were duly loaded and exposed. The photography then continued throughout the eclipse. Ten minutes later, as the Moon covered a sunspot group, clouds spontaneously formed, blotting out the Sun. The astronomers halted their observations and watched anxiously for six minutes until the clouds melted away again. They hurriedly resumed their tasks.

As more and more of the Sun disappeared behind the Moon, De la Rue noticed the azure blue of the sky giving way to an indigo tint. Around him, the landscape assumed a bronze hue. He wondered what the spectral analysis of Kirchhoff and Bunsen would reveal about this change in the Sun's rays. When the Moon reduced the Sun to a narrow crescent, he saw the shadows cast by the equipment became suddenly sharp. They reminded him of the well-defined shadows produced by electric lights. As totality finally arrived, the landscape fell into darkness and the assembled crowd suddenly hushed. Church bells tolled across the valley. De le Rue began sketching through the telescope. He could make out the sepia-tinged lunar surface that sat in front of the Sun, but his attention was drawn to the prominences of pink fire that peeped out from behind the Moon.

Finishing his first sketch, he looked up at the sky with his naked eye. Around the pale streamers of the Sun's corona, the sky was a deep indigo, giving way to a deep red and orange at the horizon. Embedded in this unusual sky were the blazing pinpricks of Jupiter and Venus. He cast his gaze around the landscape and saw the grand spectacle of the mountains turned blue with the Moon's shadow draped across them. Enthralled by the eerie beauty of the event, he regretted ever having encumbered himself with scientific duties. He mentally vowed that if another eclipse presented itself, he would become one of the crowd, de-

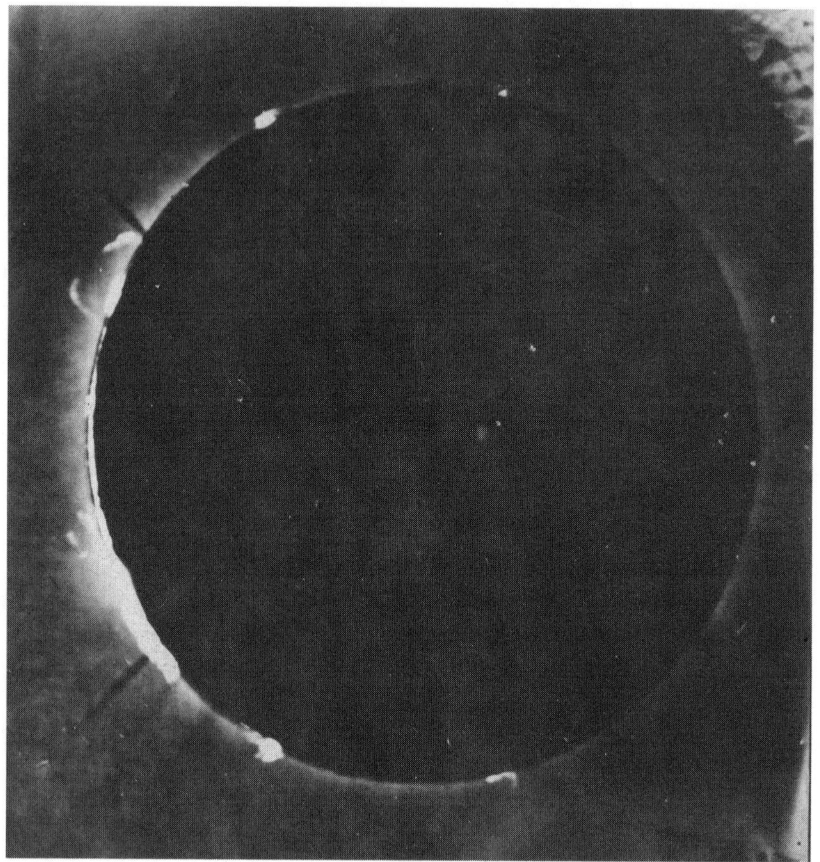

Warren de la Rue's picture of the totally eclipsed Sun. (Image: Royal Astronomical Society)

voting himself solely to wallowing in the spectacle. With this promise made, he tore his eyes from the surroundings and pressed into the telescope's eyepiece once more to resume sketching.

Around him the process of photography continued. One plate had already been exposed during totality and was being immersed in the developing liquids in the back of the tent. A second was being loaded into the photohelioscope. As De la Rue set about his second sketch, he could contain his curiosity no longer. He called to the darkroom, asking for word on progress, and was thrilled to hear a shout returned: the photography was a success; there were prominences visible on the plate. De la Rue felt a thrill of intense pleasure at the words. He and his team had done what no one had achieved before: captured the depth of an eclipse for posterity.

Enlivened by their success, the team worked on for hours, even after light returned to the land. They sketched and measured and photographed as the Moon passed completely away from the Sun. The day after, George Airy, who had been touring the various eclipse parties in Spain, arrived at Rivabellosa. He inspected the eclipse photographs with great satisfaction, praising De la Rue and his team for their efforts. On the expedition's return to England, De la Rue's triumph was celebrated throughout the land. Yet what no one realized at the time was that during the eclipse, the Sun had been showing them something extraordinary in its outer atmosphere—something that, if they had only recognized its importance, would have given them the clue to the way Carrington's flare had triggered the subsequent magnetic storm.

On the day of the 1860 eclipse, the rotation of the Earth had swept the path of totality across Canada, a tiny bit of America, over the Atlantic, through Spain and northern Africa before the celestial alignment was broken and the Moon's shadow lifted from the Earth's surface. A Mr. Gillis was one of the first to see the eclipse. He was an experienced eclipse observer for the U.S. Coast Survey and took up position at Steilacoom, near present-day Tacoma, Washington, on the Puget Sound, before dawn. As the Sun rose over his location, the Moon's shadow was already clearly visible over part of the Sun's disk.

As the Moon completed its maneuver, the glorious corona burst into sight and Gillis saw that it was particularly magnificent this time, with large fingers of ghostly white light reaching away from the black disk of the Moon. He knew he had just minutes to take in the sight and forced himself to concentrate first on the bright inner corona that girdled the Moon. There were ten or more pink flames that stood out from the white coronal light. Having observed each one, he turned his attention to the corona, tracing it away from the Sun and concentrating on the faint details. He concluded that the corona was made up of nothing but radial beams, some thick and some thinner, with dark interstices between them. All of them pointed straight outward from the Sun.

At the same time, in Ungava Bay, Labrador, astronomer R. N. Ashe was facing disappointment. The eclipse was not due to arrive for another hour, but the sky remained stubbornly filled with broken clouds.

All he could do was hope for a gap at the required time. He set his three-inch telescope at the ready and waited. As totality arrived; the world around him was plunged into darkness and he looked hopefully through the eyepiece. A gap appeared, revealing the corona as a bright halo around the Moon. As he took his first look at the spectacle, a bright flash caught his eye. In the southwestern quadrant of the corona he saw "a white flame, shooting upwards to a considerable distance." Although bright, it possessed the same milky quality as the rest of the coronal light, and gave the impression that something had been ejected from the Sun's surface. Before Ashe could see more, the clouds reasserted their authority, blocking his view. A few moments later and light returned to the land as the eclipse ended from his vantage point, leaving him to wonder what the white flash had been.

It took two hours for the inexorable turning of the Earth to align the Sun and Moon over Spain, where many European astronomers were waiting. As the temporary darkness swept across the shores of the Bay of Biscay, several observers noticed a disturbance in the southwestern corona. Instead of a straight, spokelike protrusion of coronal light, one streamer curved gently out to nearly two solar radii. Several minutes later, keen-eyed observers in the center of Spain recorded in their sketchbooks that a second strand had joined it. Originating from the same point on the solar surface but curving the opposite way, the complete shape mimicked the outline of a tulip head. When the eclipse reached the Mediterranean coast of Spain, eleven minutes had elapsed since Spanish landfall. The astronomers on this side of the country saw a new development. The tulip detached from the Sun to form a separate oval bubble in the corona itself.

The last descriptions of the eclipse were recorded in Algeria, twenty minutes after the event had been lost from sight in Spain. French army engineers at Batna and an astronomer at Lambesa remarked on the southwestern disturbance. Yet no astronomers thought to investigate what the coronal bubble represented after the eclipse. Worse still, some called it into question.

Despite a number of observers calling the bubble "spectacular," one in three eclipse watchers failed to notice it. Perhaps some were poor observers, or were using poor equipment or simply did not have time to notice it during the few brief minutes of totality. One of those who failed to see the bubble was the highly respected astronomer Father Pietro Angelo Secchi. He was in charge of the Vatican Observatory in Rome, and this gave him an elevated reputation throughout Europe. His observing skills were usually acute. He drew one of the first maps of Mars and discovered new features on the Sun's surface. Yet on the day he observed the 1860 eclipse from Desierto de las Palmas, he failed to see the bubble, and this gave the phenomenon a serious credibility problem.

Such discrepancies in the eclipse reports turned astronomers away from sketches and fostered the growing belief that only observations recorded with a camera's photograph could be trusted. Photography was fast becoming the chosen way to capture the purest essence of nature, devoid of human interpretation. De la Rue's triumphant pictures of the eclipsed Sun embodied the new thinking. Because they had only just captured the prominences and bright inner corona, rather than the fainter outer corona where the bubble was visible, the photographs could not settle the debate.

In reality, the bubble was an eruption of solar particles that sometimes takes place at the same time as a solar flare and which, when directed at Earth, collides to cause magnetic storms. But with such scant knowledge of what the solar corona was, let alone the nature of electrified solar particles, the significance of the observation escaped them. Perhaps if the bubble had been pointed to Earth and a great magnetic aurora had taken place the next night, astronomers would have paid it more attention. As it was, despite the fact that two-thirds of observers saw it, the structure was all but ignored. If it were discussed at all, its reality was always left in doubt.[3]

In the autumn of 1860, Professor James Challis, who had bungled the discovery of Neptune in 1846, quietly requested that he be allowed to relinquish his directorship of the Cambridge Observatory while retaining his professorship in astronomy, the coveted Plumian professorship. The

[3] It was not until a century later that the NASA space station Skylab took regular images of the Sun's corona and recorded images of eruptions that looked reminiscent of the 1860 sketches. This prompted J. A. Eddy of the High Altitude Observatory to reopen the debate about the reality of the 1860 event.

university convened a special syndicate to consider the request and determine how to fill the vacant directorship.

Word of this development reached Carrington in January 1861 along with a message from Challis to await further instructions as to how to apply. It had been mooted before that the observatory's director should be a separate post, rather than an additional duty to one of the astronomy professorships, which tended to be held by mathematicians whose expertise lay in theoretical astronomy. Observing was no longer a case of glancing through the telescope and speculating, it was now a precision science in its own right. As such, there was a growing divide between the skills of theoreticians and observers. Carrington had dedicated himself to becoming the best observer of his generation.

On paper, he was the perfect choice for the Cambridge position. He was a graduate of the university, had succeeded in reaching the top echelons of science, and was both a Fellow of the Royal Society and a leading member of the Royal Astronomical Society. Carrington certainly thought of himself as the natural choice and at once began to marshal allies. He wrote to John Herschel requesting patronage and, perhaps mindful of the Oxford experience, wrote to Airy in spite of their differences, and explicitly asked for his help in securing the Cambridge post. A note of desperation entered Carrington's letter as he told Airy that his hindrances were now so great that he had all but stopped astronomical work and that without the salvation of the Cambridge position, he would soon have to relinquish all hope of remaining an astronomer.

Airy refused to help, stating that the matter was none of his business. Nevertheless, he said that if it were his business, he would detach the observatory duties from the Plumian professorship and hand them to the Lowndean professor of astronomy and geometry. Carrington must have thought the Astronomer Royal's suggestion tantamount to a joke. It made no sense because the Lowndean chair had passed, in March 1859, to John Couch Adams, a theoretician. All that Cambridge would gain by such a move was a totally unsuitable observatory director.

Carrington waited for the public announcement of the position's availability, little knowing that negotiations were already taking place in Cambridge behind the closed doors. Suspiciously, the syndicate's plans matched exactly what Airy had suggested. At a meeting of the special syndicate, of which Adams was a member, the theoretician was asked to accept the observatory responsibilities in addition to his professorial duties. He was highly unenthusiastic, pointing out that he was far from ideal for the job and recommending that the university create a position and appoint a dedicated director. His syndicate partners were not impressed by his opinion and repeated their desire that he fill the role.

Feeling pressured, Adams wrote privately to the vice-chancellor of the university, setting out the reasons he believed himself unsuitable to direct the observatory. "I feel that I should be likely to contribute more to the progress of science and to the credit of the University by continuing to cultivate that branch of astronomy to which I have hitherto almost exclusively given my attention," he wrote, meaning mathematical theory. Then he reiterated his belief that the directorship should be a separate position, filled by an experienced observer.

Yet his pleas were to little avail and eventually Adams agreed to take on the task, providing that a number of extraordinary conditions were met. First, that he would never have to observe; second, that he would never have to process the observed measurements into usable data; and third, that in the event that the administrative duties associated with the post became too much, he could resign. Astonishingly, the syndicate agreed that assistants could carry out all observations and mathematical reductions under the loose management of Adams.

When Carrington heard the news, he was crushed. There was to be no public call for applicants and no new beginning for the observatory, just a quick administrative reshuffle. Carrington could hardly believe it. He wrote to Herschel in despair, telling him that his career in astronomy was over. With nothing now to lose, Carrington began a last-ditch attempt to claw his way back into professional astronomy.

On 13 April, he wrote to the vice-chancellor of the University of Cambridge, taking issue with the decision to appoint a theoretician to a practical astronomy position. His words implied heavily that over and above his own stake in the affair, such a move was little short of imbecilic. He wrote:

> The Directorship of a Public Observatory is not like the Curatorship of a Library or Museum, but is, in the Astronomical World, supposed to demand qualifications of a special kind which all theoretical men, even of the highest class, are not found to

> possess; and the experience of late years has not tended to obliterate, but rather to render still more distinct, the difference between those qualities which mark the man eminent in Theoretical Astronomy and those which mark the man capable of conducting efficiently the practical work of an Observatory. It would be easy to show that, where both are combined, a person thus doubly qualified will in the present day find it necessary to make choice between the two; and that he will speedily find that to do full justice to the Observatory and attendant correspondence, he must nearly abandon analytical research, or, on the other hand, shortly be conscious that, by exercising his appetite for investigation, the Observatory under him has begun to languish.
>
> Although I shall always be among the first to recognise the eminent abilities of Professor Adams in his own department of Astronomy, I shall, without hesitation, claim the right on both public and private grounds to contest with him the relative fitness for such an office as the Directorship of the Cambridge Observatory.

Such criticism of Adams must have been difficult for Carrington. Both men had come to know and respect the talents of the other. Adams had even been one of his proposers for Fellowship to the Royal Society. Nevertheless, Carrington went on to assert that it was an injustice for the university not to appoint the best person in the country to the job of director, and "I claim at this time to be that person." He detailed his successes and promised Cambridge a share in the glory of his sunspot catalog, which he described as "a very considerable series of observations," the results of which, "though as yet only sketched out, are admitted to be of high physical interest."

Finishing his letter by declaring his intention to publicly contest the appointment of Adams, Carrington waited for the reply. His words obviously piqued the interest of the vice-chancellor, who responded within a few days to grant that Carrington could contact members of the university's senate, who were due to consider the recommendation of the observatory syndicate in a few days' time. Carrington wrote by return post asking for names, as he had lost track of the residing senate. By the time the list appeared, it was too late. The senate had accepted Adams's appointment.

Carrington was incensed. He wrote again, protesting at the irregular way the appointment had been made. He claimed that the syndicate had exceeded its position in nominating an individual, especially one from within its own ranks, and ascertaining that person's consent to the position. What annoyed Carrington the most was that Adams had accepted an increase of just £250 in his salary for taking on the observatory. At only half the salary offered for the equivalent Oxford position the year before, Carrington believed that the Cambridge sum reflected badly on the worth of practical astronomers. Cambridge, he claimed, set the precedent for other establishments to follow, and this paltry sum was a disservice to astronomers of ability who were currently struggling to establish themselves as professional scientists. In Adams's acceptance of the appointment at the proposed salary, no matter how reluctantly, Carrington believed that the theoretician too was complicit in the retardation of professional astronomy.

A week later, Carrington fired off another broadside, repeating his previous points. No longer able to ignore the criticisms, which were circulating widely among the dons in Cambridge, the senate sat to consider the protest. After due deliberation, they upheld both the syndicate's conduct and Adams's appointment. They did not consider the position a "public" appointment, as Carrington had referred to it, because it was neither a government-funded job nor royal appointment. In other words, they could appoint whom they wanted, how they wanted. The matter was closed.

Carrington took his last sunspot observation on 24 March 1861, ending the grand study he had hoped to conduct for a full eleven-year solar cycle, after just over seven. Doggedly, he began preparing his house and observatory equipment for sale, determined to make good his threats to the world and quit astronomy for good. With his work abandoned, the sunspot data would lay trapped inside his notebooks, effectively lost to the world.

At one o'clock on 17 July, almost a year to the day since the Spanish eclipse, a small group of interested parties gathered at Garraway's Coffee House, Change Alley, Cornhill, and bid for the Redhill observatory equipment. Lot by lot his observatory was carved up and sold off. Carrington watched the crowd bid for his beautiful brass equatorial telescope, the one he had seen the solar flare with. However, the most painful loss was the sale of his second telescope, the transit circle, with which he had measured the stars for his Gold Medal–winning Redhill catalog. The Radcliffe Observatory, Oxford, bought it for £420. The very observatory that had helped to destroy his career by not employing him now took one of his telescopes.[4]

After the sale, the observatory wing at Redhill became a hollow drum. Carrington's dream of spending his life in the pursuit of astronomy had slipped away. He had established himself as one of the greatest living astronomers and yet his expertise had not protected him from the twists of fate. His undoubted knowledge had been insufficient to secure him a professional position. Instead, petty politics, favors, and his own fiery impatience had conspired against him. In just two years, everything he had fought to achieve during the previous decade had faded away, just as the brilliant solar flare had died before his eyes.

With his observing equipment gone, there was no need for him to remain at Redhill. He sold his house and moved closer to the hated brewery, so that he could attempt to settle into the life of a businessman.

[4] The telescope continued to be used at the Radcliffe Observatory for professional work into the twentieth century. It is now on display in the Museum of the History of Science, Oxford, U.K.

NINE

Death at the Devil's Jumps, 1862–1875

Unable to make any headway in their investigation of the links between sunspots, solar flares, the daily variation of compass needles, and the frequency of magnetic storms, a number of astronomers began to lose faith in the connection. At the forefront of the doubters was George Airy. He was probably swayed from giving credence to the link by his personal enmity toward Sabine as well.

On 23 April 1863, Airy presented an analysis of twenty years of Greenwich magnetic data to the Royal Society. He had carefully kept this data within the confines of Greenwich so that it did not fall into the clutches of Sabine. First, Airy talked about the daily drift in the magnetic needles. He reported a general strengthening of the variation from 1841 to 1848, a decrease from 1848 to 1857, followed by another increase. He mumbled some comments about this probably reflecting the general magnetic condition of the Sun but failed to mention that his readings fell into agreement with the sunspot cycle, which had reached a maximum in 1848 and minimum in 1856. Thus he overlooked an opportunity to publicly confirm the validity of Sabine's sunspot–magnetic variation.

Next he turned to an analysis of the magnetic storms recorded at Greenwich. This time he did refer to the solar cycle but proclaimed that he saw no evidence for it in the frequency of the magnetic storms. He later proposed that, if any periodicity did appear in the data, it was closer to a six-year cycle. Airy's big mistake in making this statement was to fail to consult the sunspot catalogs compiled by Schwabe, Spörer, or Carrington. If he had, he would have discovered a peculiar feature of the solar cycle that continues to this day. It is that, even though the average number of sunspots begins to decline, anomalously large and active spots often appear several years after solar maximum, temporarily raising the number of magnetic storms on Earth. For example, the extraordinary Halloween storms of 2003 140 years later took place in the declining phase of the solar cycle. If Airy had looked at the sunspot data, he would have seen that the brief return of the magnetic storms corresponded perfectly with the temporary appearance of giant sunspots, thus strengthening Sabine's assertion that the storms and sunspots walked in lockstep.

Airy was also doubtful that the currents experienced on the telegraph lines were induced by the aurora. Instead, he wondered whether the Earth was generating these currents. As they flowed across its surface and through the conveniently placed telegraph wires, did they force the auroras to shine overhead? He organized two experimental telegraph lines to be erected from Greenwich. One stretched ten miles east to Dartford, the other eight miles south to Croydon. Scientific equipment that would continuously measure any current flowing along the wires was connected.

However, Airy made an enormous mistake. He grounded the telegraph lines by connecting them to water pipes at the railway stations. These water pipes acted as antennas themselves, more powerful than the telegraph lines they were supposed to serve. The resulting data were therefore awash with false readings and, unsurprisingly, failed to correlate to the appearance of auroras or the movement of the Greenwich magnetic needles. Unaware of the error in the experimental setup, Airy used the mismatching results to promote skepticism in the Sun as the ultimate source of magnetic storms, auroras, and Earth currents. To some, it began to look as if Carrington's solar flare had been a red herring.

As for Carrington himself, if he had thought that he could exorcise the astronomer within by simply selling off his observatory, he had seriously underestimated the scale of his own possession. Now living closer to central London, he found it easier to attend meetings of the Royal Astronomical Society and the Royal Society. Hearing updates on the solar work at Kew no doubt intensified his feelings of falling behind the onward march of astronomy. Predictably, the sunspot data began to call to him, exerting its irresistible siren song even in the face of his daily business commitments. If his seven years of effort were going to mean anything, he had to publish the results as soon as possible, and that meant countless hours of further work to turn the observations into meaningful data.

Unable to let his dream of a grand solar catalog die, he began snatching time from the brewery and working late into the night, preparing his work for publication. By 1863, it was ready. All Carrington needed was to find a sponsor willing to bear the cost of printing the tome. In 1857, the Admiralty had paid for the printing of his star catalog as an aid to navigation. But the sunspot work was pure science and that might make finding someone willing to foot the bill more difficult. In fact, none other than the Royal Society agreed to bear the costs. It recognized the world-class quality of the work, even if it had fallen short of Carrington's original intention to collect data for a full eleven-year cycle. Soon after publication, Balfour Stewart used it to show that auroras appeared only when large sunspots were visible on the Sun's face.

Unfortunately, Carrington's satisfaction at his achievement was short lived. With the publication of the sunspot catalog, his last ties to active astronomy were severed. Without an observatory he could perform no observations, and without data he could not even bury himself in astronomical calculation. The large volume, however handsome, must have seemed like the tombstone at the end of his career rather than the crowning glory he had once envisaged. He began to dream of a voyage to Chile, emulating John Herschel's South African sojourn. In Herschel's case, the massive catalog of nebulas he recorded had catapulted him to preeminent scientific status. Carrington dreamed of observing the southern stars, in a mirror image of his northern star catalog that had been so well received.

Before Carrington could advance his plans, calamity struck. He fell gravely ill. The nature of his malady was never explicitly disclosed but could have been a nervous breakdown or most likely a stroke. He became so weak that he was confined to his house for months. Carrington's recovery was so slow that he feared his life would never return to normal. Festering in bitterness, he communicated with the Royal Astronomical

Society by letter only, at one point accusing the society's treasurer of incompetence in his presentation of the accounts. Although Carrington did not accuse the treasurer of dishonesty, he did ask a number of searching questions about the finances. Eventually, Carrington demanded a vote of confidence in the accounts. When George Airy and the other officers of the society refused to accede to this outrageous demand, Carrington published a pamphlet at his own expense on how he thought the society's accounts should be presented. He titled it "Revenue Account versus Cash Account—A Breeze." Where once he had been the staunchest officer of the Society, he had now become a thorn in its side.

As he recovered some measure of physical strength, Carrington finally found the courage to sell the hated business. Although he was just thirty-nine, he determined to retire on the proceeds and resume astronomy. As he had done a decade earlier, he began searching for a site on which to establish a home and an observatory. He struck upon a plot in the village of Churt, Surrey. It contained a conical hill some sixty feet high, known as the Middle Devil's Jump. Someone had excavated a tunnel that led into the center of the hill, possibly for a grotto that was never completed. Carrington imagined establishing an observatory underground, where temperatures would be stable. The telescope would peep up just above ground level, saving him the expense of constructing a new building for it. He bought the land and arranged for the construction of a large house at the foot of the hill, within clear view of the local inn, known as the Devil's Jumps.

Carrington was welcomed into the village and soon found himself manager of the Church of England school there. His first years at Churt were spent living the quiet life of a bachelor, puttering around his troglodyte observatory and contributing a few minor observations to the Royal Astronomical Society. Some of his neighbors eventually became suspicious of him, and village gossip started to circulate. The rumors suggested that Carrington was constructing a glass-fronted coffin for doubtlessly nefarious reasons. The reality was that he was creating a large clock for use in his observatory, and what the locals had witnessed was the delivery of the pendulum case.

Once again his life changed dramatically during 1868. Carrington was in London, strolling along Regent Street in the late afternoon when he fell in love at first sight. He approached a strikingly beautiful woman and engaged her in conversation. He learned that her name was Rosa Helen Rodway and that she lived with her brother, William, in a rented house in Cleveland Street. Emboldened by her unmarried status, Carrington persuaded her to accompany him that night to the theater. At the end of the night, he arranged to meet her again.

Rosa was a poorly educated woman. She could read but few words and the art of writing escaped her completely. She seemed an unlikely partner for a Cambridge graduate, and yet, despite their educational gulf, the two began a courtship. During this time, Rosa continued to live with her brother. Carrington would sometimes visit and, in so doing, made the acquaintance of William. En route to one such rendezvous, Carrington obtained a special license to marry and asked Rosa to be his wife. She refused his offer but continued to see him. So Carrington bided his time and continued to woo her. In the summer of 1869, he again arrived at the Rodway home with a marriage license. This time he brought and read his will, making it clear that his personal wealth was the considerable sum of £25,000. He promised to make ample provision for Rosa in this will and asked her once more to marry him. She accepted.

On 16 August 1869, they were married. On the marriage certificate, Carrington entered his occupation as simply "gentleman" and Rosa signed her name with a cross. After the ceremony, the honeymooners left for Paris. Upon their return, Rosa refused to live with Carrington in Churt, claiming it was inappropriate for her to do so until she had obtained sufficient education to fulfill her role as the wife of a gentleman. Instead, she wished to take a house in Battersea with her brother. Carrington paid her rent and returned to Churt, visiting occasionally to exercise his conjugal rights.

After nearly two years of this disjointed marriage, Carrington's patience, which had never been in full supply at the best of times, was finally exhausted. He insisted that Rosa leave her brother and live in Churt, regardless of her lack of education. As an incentive, he refused to stump up any further rent. Rosa moved to Churt, giving the locals something new to gossip about. Carrington was frequently away from home, sometimes for days on end, and local tongues began to wag with talk of a handsome man seen skulking about the house when the master was away.

On Saturday 19 August 1871, sometime around midday, the hall bell at the Carrington residence rang. Rosa was sitting in the front room and intercepted the household servant in the hall, saying that she would answer the door herself. Upon finding William Rodway standing there, Rosa withdrew and attempted to close the door from the hall into the kitchen, but it caught on the coarse cocoa mat and remained ajar. Sensing mischief afoot, the cook took the opportunity to spy through the crack, the servant hovering behind her. Doubtlessly both were hoping to learn something to contribute to the local gossip. They recognized the man at once, having seen him prowling the hill earlier that morning.

Rosa stepped out onto the front step and drew the door up behind her. Rodway made circles in the ground with a walking stick. He asked Rosa to return a cloak, a shawl, a small dog and £3 that she owed him. She asked him to wait in the nearby Devil's Jump Inn and she would have the shawl and cloak brought to him. The dog, however, would have to wait until the return of her husband because Rodway had given the animal to him, not her, shortly after the marriage. Drawing her attention to the dirt that Rodway was stirring up with his stick, he said, "I rose you from that." Then he turned to leave. Rosa watched him reach the corner of the house and turn around. Striding back he accused her, "You have been a bad woman to me," and raised his arm as if to strike her. That was when Rosa saw the knife he was holding, its five inches of gleaming blade falling toward her heart.

She had time only to raise her left arm in defense. The knife struck with tremendous force, pushing clean through the flesh of her forearm and into her chest between her fourth and fifth rib. Rodway pulled out the weapon as Rosa stumbled back into the house, her black silk dress slicked with blood. Screaming, she turned into the hall. Rodway struck again, this time landing the knife deeply in her back. The blow sent her sprawling to the floor near the hall stove. She twisted around to face her assailant.

Rodway loomed and kneeled over her. He reached down toward her bloodstained chest and grasped a fistful of her dress. He lifted her torso from the floor and reached around to pull the knife from her back. She pitifully asked for his forgiveness. "I will, I will and may God bless you," he said. Straddling her, he raised the knife once more. Rosa cried out to the cook, whom she saw watching from the crack in the kitchen door but the woman did nothing except gawk in fright. The servant ran to raise the alarm.

In desperation, Rosa grabbed at the blade, cutting her fingers. She transferred her grip to Rodway's bushy whiskers and pulled. Amazingly, when he next brought the blade to bear, it was on himself. He made stabbing motions toward his own chest six or seven time, adding his own blood to the hall floor.

"Don't! Don't! Throw the knife away," Rosa cried.

Somehow her words penetrated his frenzied state and he cast the weapon aside. Rosa seized her chance, wriggled out from under him, and made off through the front door. Dazed by the attack and lightheaded from the blood loss, she stumbled toward the inn. Rodway caught up with her moments later and grabbed her hand. "It will all be over soon, we shall meet again in heaven," he told her.

This gruesome scene, played out in full view of the inn, brought people running. Several took charge of Rosa and helped her to the inn. One witness ran to his horse and cart to drive the six miles to Farnham to fetch a doctor. Another went for the police. A naval pensioner apprehended Rodway with a sharp, "Halt!" Rodway offered no resistance, saying, "I have done all there is to be done and if I have injured Rosa then I am very sorry for it." As the retired seaman led him to the inn, Rodway claimed that he had intended to commit suicide in front of Rosa.

Inside, Rodway asked for a pen, ink, and paper. He quietly scratched out a faltering letter, sealed it in an envelope, and handed it to the landlord with a penny for postage. Shortly afterward, the police arrived and arrested Rodway. He appeared before the Farnham magistrate on Monday morning. In a packed courthouse, the description of his crime was read out. Rodway sobbed and sighed at intervals throughout. The court was told that Mrs. Carrington was too ill to attend. Rodway was remanded for a week and the police began investigating. By the week after, when Rodway again appeared before the magistrate, the police had obtained one undeniable piece of information. Rodway was not Rosa's brother. Carrington himself sat impassively in the courtroom as he heard that the fifty-two-year-old Rodway, a former Dragoon Guard, had been on intimate terms with the twenty-six-year-old Rosa for some years. It was stated that Rodway had given his consent for her to marry Carrington only after the astronomer had offered him £2,000. Then, in a fit of romantic jealousy, Rodway had tried to murder her.

The village gossips loved this. It meant Rosa had lied to Carrington, calling Rodway her brother to mask the fact that she was living with him as his common-law wife. Even after Carrington and Rosa had been married, she had continued the deception. When she had insisted on returning to her brother to seek an education, she had been cuckolding Carrington, using his fortune to keep Rodway as well, who had given up work to live off the spoils of Rosa's marriage.

But Carrington had a secret of his own. He had uncovered the true situation shortly before he had forced Rosa to live with him in Churt. By relocating her to the rural wilderness, he must have hoped that he could break the attachment she felt to the handsome Rodway and bury the scandal once and for all. Now, this sordid business was going to be laid bare in a public trial.

The case was heard months later at the spring assizes in March 1872. Carrington stoicly listened to the testimonies but offered none himself. Rodway was described as "a fine tall, powerfully intelligent looking man." He revealed how he had met Rosa in Bristol in 1865. He had been working at that time in the Tom Thumb circus. The two had become lovers and moved to London, where Rodway had become a publican. Then she had met Carrington and the relationship became a triangle. According to Rodway, Rosa was at best confused about which man she wanted to be with and, at worst, entirely duplicitous toward them both.

On the eve of her departure to Churt, Rodway had handed her some addressed envelopes and they had agreed on a code. The illiterate woman would mark the note with crosses if she were coming to London and with dots if Rodway were to visit Churt. He described how they had used the system to continue their affair. At times, the despair of losing her to Carrington had made Rodway threaten to take jobs overseas to see whether the prospect of his prolonged absence would win her back. When this had not worked he resolved to commit suicide in front of her. He entered Joseph Moreton's Oxford Street shop, in London, and asked to see knives. Rejecting the first one proffered as too small, he pored over the other weapons until he chose one with a spring-loaded five-inch blade and a cross-catch between the blade and the handle, allowing it to be used as a stabbing dagger.

His explanation for Rosa's terrible injuries was that she had intervened in his attempt on his own life. As he brandished the blade against himself, she had been struck accidentally. The wound in her back had occurred by her falling onto the knife when she tumbled in the hallway.

Rodway wept repeatedly during Rosa's stout rejection of these claims. But it was the bloodstained letter he had written in the inn during the aftermath of the crime that condemned him. It read, "I have stabbed the woman to the heart, I hope." Rodway's legal adviser argued that the sentence was incomplete and that the weakened Rodway had intended to write, "I hope I am misinformed." Nevertheless, when the jury retired to consider its verdict, the deliberations took just five minutes. Rodway was found guilty of wounding with intent to murder.

Before he could determine on the sentence, the judge asked the jury to consider a final piece of evidence. He called Edwin Gazzard, a former prison warden, to the witness stand. Gazzard claimed to have met Rodway almost twenty-three years previously. At the time, Rodway had been known as Edward Smith and was enduring a year's hard labor for the manslaughter of one Rebecca Gill. Rodway instantly denied the charge and the jury were instructed to decide the matter. Was William Rodway really Edward Smith? If so, the sentence would undoubtedly be death.

This time the jury rejected the claim on the grounds of insufficient evidence. The judge told Rodway that he had nevertheless been convicted of a very serious crime. Had he killed Rosa, as was clearly his intention, he would have hanged. Her life had been saved by nothing but providence, and to mark the enormity of Rodway's crime, he was to be kept in penal servitude for twenty years. Rodway listened to the verdict with calm dignity and was then led away. He died a few years later, still in custody.

Following the trial, Carrington attempted to put the personal drama behind him and to live a quiet life with his wife. In January 1873, less than a year later, he signed a new will leaving everything to Rosa. Periodically, he would journey to London to attend scientific meetings and catch up on the latest developments. The Sun was now the single most observed object in the heavens. Predictably, discoveries followed. One of them struck a blow against George Airy's public skepticism about the Sun's role in magnetic storms. People had always wondered why more flares like Carrington's weren't being seen around the times of intense magnetic activity. Now they had an answer.

The answer came from astronomers using the spectral analysis methods pioneered by Bunsen and Kirchhoff. It was that flares were indeed taking place but, unless they reached exceptional proportions, they remained below the threshold of detection by the eye. But with spectroscopes, they were clearly revealed. The flares showed up when dark Fraunhofer lines of absorbing gas changed to bright emission lines, indicating that the gas was suddenly being heated to incandescent temperatures. The more astronomers studied the solar spectrum, the more of these inversions they saw. Usually these spectroscopic flares took place above sunspots, just like Carrington's "white light" flare, and usually they lasted a few minutes, the same duration as Carrington's flare. When compared with the occurrence of magnetic storms and auroras on Earth, there seemed to be a broad coincidence with the spectroscopic flares.

More and more astronomers began to believe that Carrington's flare had been a particularly exceptional example of the phenomenon that precipitated magnetic storms and auroras. However, Airy's skepticism remained, largely because the details of what caused solar flares, how the magnetism communicated itself across space, and why the attendant magnetic storm and aurora lagged behind the flare by many hours completely eluded them. Sadly, these final answers did not come quickly enough for Carrington. During 1875, tragedy visited him once more.

In the aftermath of Rodway's attack, Rosa had been left severely traumatized. She had been prescribed the sedative chloral of hydrate, which she took at night to ease her to sleep. Carrington too, it seems, became addicted to the drug.[1] On 17 November, Rosa failed to wake. It was obvious at once that an overdose of the drug had killed her. Beyond a few grams, the sedative, which acts directly on the nerves of the brain, in-

[1] In later years of the nineteenth century, chloral of hydrate was mixed with alcohol and known as a Mickey Finn, supposedly after a Chicago taverner who surreptitiously drugged his clients before having them robbed.

duces paralysis of the heart and lungs. An inquest was held at the Devil's Jump Inn and the coroner confirmed that the cause of death was suffocation. He delivered a stern reprimand to Carrington for not providing the correct medical supervision for his wife. This inevitably sowed the seeds for yet more rumor and gossip as villagers wondered whether Carrington had deliberately overdosed Rosa.

Carrington fled the scene. He dismissed his servants and headed out of Churt. He was seen arriving in Brighton where his widowed mother lived. A week later, he was spotted returning to his empty house. When several days elapsed with no further activity from the Carrington household, the locals became concerned. The police were called and two constables forced their way in. At first, it seemed that the house was empty but, as the search continued, one officer found a locked door in the servant's quarters. Unable to obtain an answer, they forced that door too. Carrington was inside, dead.

He lay on a mattress that he had evidently dragged from the bed and placed between the bed frame and the fireplace. His back was turned to the ashes of a spent fire and a handkerchief was knotted around his head. Upon examination, it was found to be a poultice made of tea leaves that Carrington had strapped to his left ear before collapsing on the mattress. Empty bottles of the drug that had killed Rosa littered the house.

The coroner returned a verdict of death by natural causes, citing a probable brain hemorrhage. But as news of Carrington's death circulated, so too did suggestions of a suicide, brought on by the guilt of having murdered his unfaithful wife.

The Royal Astronomical Society gave him a full and generous obituary, glossing over the skirmishes he had fought within various academic circles, the Society itself included. The Royal Society, however, allowed his death to pass unmarked, even though he had bequeathed both learned organizations £2,000 each.

Carrington's last will, signed just over two years before his death, requested that he be interred in an unmarked grave on his estate at the Devil's Jump. No service was to be read and the total cost of the proceedings should not exceed five pounds. He also specified that the grave should be between ten and twelve feet deep. This was an understandable precaution at the time against grave robbers. Another specification reveals that, although Carrington had maintained a lifelong distance from religious belief, he did have some supernatural thoughts after all. He requested that neither his chin be shaved nor his shirt be changed following his death. It was commonly thought that witches could take possession of a soul and divert it from its path to heaven if they had access to hair or bodily secretions from a recently deceased person. Carrington was clearly hoping to avoid any possibility of that.

His mother chose to ignore his request to be buried in the mound of his observatory. Both he and Rosa were laid to rest in the family grave at West Norwood Cemetery, Greater London. On the headstone was chiseled the Latin inscription "Sic Itur Ad Astra," meaning "thus do we reach the stars."

Thankfully for astronomy, the academic heir to Carrington was already quietly at work, collecting a mountain of his own sunspot data that he would eventually sculpt into a scientific edifice that no one could demolish.

His heir was Edward Walter Maunder.

TEN

The Sun's Librarian, 1872–1892

Three years before Carrington's death, twenty-one-year-old Edward Walter Maunder was working in a London bank. In late 1872, he saw an advertisement for an assistant's job at the Royal Observatory. Instead of the usual prerequisite degree, which Maunder did not possess, all that was required was for him to sit for a London Commissioners' test. This was the first test of its kind and was held in an attempt to reform the civil service by appointing jobs to those capable of performing them, rather than to those with formal qualifications. In this way, politicians hoped the positions would be more equitably appointed, because a university education was still largely the preserve of the privileged.

It was the perfect chance for Maunder. He possessed a restless curiosity that had never been fully developed through formal education. As a youth, he had seen something that fired his imagination for astronomy, especially solar astronomy. In February 1866, as he was returning home from school, the Sun was hanging low in the west, its heavy orb partly obscured by mist. The fourteen-year-old looked on in fascination. There, on the red surface, was a plainly visible black spot. He thought it resembled the head of a nail that had been driven into the Sun.[1]

Maunder was astonished by the sunspot, it was the first time he had ever seen a feature on the Sun. He kept watching and several days later the conditions were repeated. As the Sun sank behind the mist before disappearing altogether below the horizon, he saw the sunspot again. It had moved across the surface. Eager to track its progress, Maunder waited again for the right observing conditions, but the next time they

[1] In John Milton's *Paradise Lost*, the author described the landing of Satan on the Sun as creating a mark like a sunspot seen through a "Glaz'd Optik Tube" (a telescope).

occurred, two or three days later, the sunspot had disappeared, carried from view by the inexorable rotation of the Sun.

The Maunders were a relatively poor but profoundly religious family. Walter's father was a minister for the Wesleyan Society, a form of Methodist Christianity that believes there should be no prejudice between members of different classes, races or sexes. When Walter was struck by a debilitating illness, his parents had little recourse but to pray. They had already lost one son and must have feared that Walter's life would be similarly short.

Thankfully, the family was spared another bereavement; Walter pulled through and began a long fight for renewed fitness. While his illness and recuperation prevented him from attending school, the restless Maunder measured the streets of Croydon by pacing them out, and estimated the angle each road made with the other by eye. He brought the figures home and drew out a scale map of the suburb where they lived.

By January 1871, Maunder's interest in science drew him to enroll at King's College, London, to study chemistry, mathematics, and natural philosophy (which would soon be renamed physics). Founded in 1829 by King George IV and his prime minister, the Duke of Wellington, King's College offered higher education to those excluded by the Cambridge-Oxford axis of exclusivity. King's College would educate women and working men through evening classes if they needed to simultaneously earn a living.

Shortly after Maunder enrolled, John Herschel died at Hawkhurst, Kent, aged seventy-nine. *The Times* lamented the passing of "one of European science's most illustrious members." Those who had worked with him at the Royal Society knew that the truth was stronger even than this. The society's obituary read, "British science has sustained a loss greater than any which it has suffered since the death of Newton, and one not likely to be replaced." It went on to praise the influence of his teaching in awakening the public to the power and beauty of science, also his role in stimulating and guiding the academic pursuit of science.

Unlike the quiet interment of Richard Carrington, Herschel was buried with a full choral ceremony at Westminster Abbey. Friends, family, and mourners from the general public thronged the cathedral where, after due pomp, Herschel was laid to rest next to the body of Isaac Newton.

Herschel departed this world at a time of deep crisis for British science. Many believed in the need for a new coordinating force, not just in astronomy but also throughout British science. The government, they claimed, was neglecting its duty to support scientists. The malcontents looked with envious eyes over to Europe, where state-funded observatories were embracing the new astronomy of spectral analysis and using it to investigate the physical nature of the celestial objects. It was clear that this was the way forward for the science. Traditional astronomy would increasingly give way to "the Physics of Astronomy," as the new techniques were being called.

As the voices of dissent rose in volume, the oligarchs of British science began to feel the pressure of younger reformers. Edward Sabine, whose magnetic crusade of the 1840s had shown the Sun's magnetic influence over Earth, was now fighting for his professional life. Now an octogenarian, his ten-year presidency of the Royal Society was ending in acrimony as he was accused of neglecting the natural science in favor of the physical ones. Unable to withstand the pressure that once he would have so nimbly sidestepped, Sabine resigned and retired.

Before the revolution spiraled out of control, the government felt compelled to act. They appointed a special commission to investigate the state of British science and recommend how it could be improved. Chaired by William Cavendish, the 7th Duke of Devonshire, the Devonshire Commission began hearing deputations. Some in the Royal Astronomical Society saw it as the opportunity to change the face of British astronomy forever.

Colonel Alexander Strange was a retired army officer from the Royal Engineers and Royal Artillery. He was a passionate advocate of the new astronomy and argued for the inauguration of a new, nationally funded astrophysical laboratory. Strange carried within him a simmering resentment of the way British astronomy was being conducted, especially the stranglehold of the Astronomer Royal that reached out from Greenwich. If Strange could persuade the Devonshire committee that a new observatory, unconnected with Greenwich, was essential to resuscitate British astronomy, he could break the Astronomer Royal's grip once and for all. But to do that, he needed other astronomers to publicly agree with him and that was going to be difficult to orchestrate.

Airy had been Astronomer Royal for nearly forty years and in that

The Astronomer Royal (1835–1881), George Biddell Airy. (Image: Royal Astronomical Society)

time he had been a loyal public servant who was trusted by the government to advise on many scientific and engineering matters. He had sat on committees that discussed such diverse topics as the gauge of the nation's railway lines and how to make the Westminster clock of Big Ben run accurately. Underscoring his stature, Airy received a knighthood during the very year that Strange chose to launch his attack.

On the eve of Strange's testimony to the Devonshire Commission, the retired colonel marched into the lion's den of the Royal Astronomical Society and attempted to rally support for his controversial views. He figured that there had never been a better time to call sympathetic Fellows to arms. Airy was an aging force, as was his loyal rear guard. Interest in the new technologies and techniques of astrophysics had swollen the ranks of the RAS with those hungry to understand the celestial objects. During the 1860s alone, the number of Fellows had risen from 380 to 509. Strange attempted to marshal these young guns into a new force by provocatively titling his presentation "On the Insufficiency of Existing National Observatories." The core of his argument was that Greenwich was already fully employed with its traditional programs and so could not be expected to efficiently add new ones. Airy had recently recommended the establishment of a separate observatory charged solely with the observation of Jupiter's moons, and Strange took this as an indication that the Astronomer Royal himself felt overstretched.

Although Strange avoided any direct criticism of Airy, the clear implication of his words was that Airy's staunch adherence to the original goals of astronomy had allowed the continental astronomers to leapfrog Britain in their achievements. Instead of inspiring the next generation of astronomers to push the boundaries of their science, Airy had become a dead hand at the tiller, anchoring British astronomy in the bedrock of a centuries-old endeavor to assist navigation through the observation of the stars. The tiny refinements this work now brought to the art of navigation were all but meaningless. Nevertheless, their prosecution still dominated the work at Greenwich.

It was not that Airy failed to understand the implications of the new science but rather that he saw himself first and foremost as a trusted public servant rather than a pioneer of astronomy. Strange considered Airy more interested in maintaining the routine of the observatory and meeting publishing schedules than in exploring the Universe. Understanding the Sun lay at the heart of astrophysical inquiry, according to Strange, and Greenwich was not the natural seat for this endeavor because how could the investigation of unknown realms be time-tabled? Who knew when the crucial insight might be observed? As with Sabine's discovery that magnetic storms rose and fell with the solar cycle, it might take decades of data gathering before the mathematicians could spot anything useful. He pointed out that the question of how the variability of the sunspots might be affecting the weather remained unanswered and, in an echo of William Herschel's justification of solar investigations, wondered how anything could be more relevant to the inhabitants of the world than understanding the Sun, the ultimate source of life and energy on Earth?

Placing this noble inquiry on an urgent footing was that Warren de la Rue had recently announced his impending retirement. He had superintended the Kew photohelioscope for almost fifteen years, including taking sole charge of its use when insufficient staff levels at Kew had jeopardized the program. During 1861, he had been forced to remove the photohelioscope from Kew altogether, erecting it in his private observatory in Cranford, where he could more conveniently fit the necessary photography around his commercial commitments. A year later, he returned the telescope to Kew, where the staffing problem was temporarily solved by one of the observatory's assistants, who deputized his own daughter to take the photographs. The young lady, Miss Beckly, proved to be so diligent an observer that an anonymous citation of her services appeared in the *Monthly Notices* of the Royal Astronomical Society:

> During the day she watches for opportunities of photographing the Sun with a patience for which her sex is distinguished, and she never lets an opportunity escape her. It is extraordinary that even on very cloudy days, between gaps of cloud, when it would be imagined that it was almost impossible to get a photograph, yet there is always a record at Kew.

The retirement of De la Rue from solar photography in 1872 coincided with a change in the administration at Kew. The Royal Society took the reins from the British Association for the Advancement of Science and, with the departure of De la Rue, solar photography was expected to cease altogether. The folly of this abandonment incensed Strange. He traced the continuous scrutiny of the Sun from Schwabe, through Carrington to De la Rue. Scarcely could he believe that the Royal Society was prepared to let a fifty-year legacy of continuous monitoring die. He told the fellows of the RAS that it was deplorable and that no time should be lost in averting "such an evil." According to Strange, this was exactly the kind of circumstance that a national astrophysics laboratory, with a defined research agenda, would avoid.

The floor erupted in debate. Airy rose to counter immediately. First, he reminded the audience that the public was reluctant to spend money on scientific endeavors. Next, he took Strange to task for mentioning the supposed links between the Sun and the Earth. Despite the growing evidence, Airy could not bring himself to believe in any link beyond mere sunlight between the Sun and Earth. He claimed that he could see no justification in "groping about for causes." He ridiculed the very idea of a nationally funded astrophysical observatory, saying that "it was the place of government not to establish philosophical institutions, but working bodies."

What Strange did not know was that Airy held a trump card; he had already opened negotiations with De la Rue and the Kew Observatory to have the photoheliocope transferred to Greenwich, where the daily photography could continue. The routine collection of data was a task that Airy felt entirely comfortable with. It had clearly defined goals that could be audited by the Treasury and gauged against value for money. He had also privately solicited advice about instituting spectroscopic work at Greenwich.

The two men next took their fight to the Devonshire Commission. Airy argued for the new work to be carried out at Greenwich, although this would require more staff and involve him in considerably more work. He saw no reason to change the way that speculative science—for that was how he saw the new astronomy—was performed. It should be left to gentlemen of passion and means. He argued that the present system allowed any man of sufficient drive to devote himself to science. Once they had proven their own worth and that of the science they pursued, they would be supported with small grants from the Royal Society and other bodies or patrons. One such example was William Huggins, a man of such drive that, using government money, the Royal Society had honored him with a specially built, high-quality telescope with which he could privately pursue a research agenda using the new science of spectral analysis. Huggins routinely disseminated what he learned at the meetings of the RAS and in the pages of its journal.

Strange decried Airy's attitude, believing it to exemplify the elitism that Victorian Britain was supposed to be eschewing. By all means leave the liberated individuals to their scientific pleasures, but do not rely on them. Without proper training and defined research goals, astronomy would meander to the whim of individuals. Even when someone did perform world-beating science, the knowledge of how to do this would be generally lost as the individual's interests changed or he died. Long-term, large-scale science was needed to restore Britain's preeminence in astronomy. As far as Strange was concerned, that meant the establishment of a national astrophysics laboratory. Strange saw Huggins not as a hero but as a marked villain. Huggins was a man of clear talent in the new realm of spectral analysis. Yet, instead of using his expertise and government-provided equipment to study the nature of the Sun and its links to the Earth's climate, he frittered it on studying the spectral differences between the Sun and other stars. To Strange, this was a blatant dereliction of moral duty. Having spent much of his career serving in India, where he spearheaded the measurement of the country's longitude, Strange had firsthand experience of the horrors that weather could inflict. If there were even the slightest chance that a monsoon drought could be predicted by studying sunspots and emergency provisions stockpiled to alleviate the suffering, then Strange felt it was government money well spent and all who received it should be compelled to investigate the connection.

The nascent revolution rumbled on. By May, part of the RAS had splintered into a minority of supporters for Strange's brave new world of breakaway astrophysical observatories. The majority favored extending the capabilities of Greenwich to include astrophysical observations and continuing the grant system for individuals who had proved themselves worthy of support. Under this weight of opinion, fostered by Airy's revelation of the plans he had already set in motion to bring both solar photography and spectroscopy to Greenwich, the RAS put its weight behind the Astronomer Royal, and Strange's coup was suddenly defeated.

Now all Airy needed was more staff to turn these programs into reality and that's where Maunder came in. Although Maunder failed his first test in 1872, coming third for two positions, he took the test again the following year and was appointed photographic and spectroscopic assistant on 6 November 1873, with a principal responsibility for solar astronomy. But upon his arrival at Greenwich, the young man found it difficult to fit in.

Airy had molded his workforce into a slick machine. For their efforts, he made sure that they were comparatively well rewarded. Back in 1826, when Airy had accepted the Lucasian Chair at Cambridge, this highly prestigious professorship had been paid at just £99 per year, the equivalent of a bank clerk's wages. In his time as Astronomer Royal, he elevated the wages of his assistants to several hundred pounds a year, roughly twice what they would have received from university observatories. Their working day lasted officially five hours; they received over a month's holiday and a pension at sixty-five. No wonder, then, that, as Maunder joined the staff, the longest serving of the Greenwich warrant assistants had been there for thirty-seven years and the shortest for fifteen years. In return for these generous working conditions, Airy expected things to be done his way, to the letter. In Walter Maunder, Airy did not quite get what he bargained for.

When vacancies had occurred in the past, Airy had handpicked the replacement and, as Carrington had so painfully experienced, also helped others in power choose personnel for university observatories. Airy clearly resented having an employee, especially one without a degree, forced upon him by this newfangled examination system. He described Maunder in one letter as "the veriest dummy I ever saw—the wrong man was chosen."

For his own part, Maunder was clearly terrified of the curmudgeonly Astronomer Royal. In his seventies, slightly stooped, and with wire-framed glasses pushed firmly to his eyes, Airy prowled the observatory. Draped in a black double-breasted frock coat, with upturned shirt collar held in place by a taut neck stock, he dispensed judgment on his workers wherever he went. On one occasion, he crossed Maunder's path. Even though Maunder had been working at the observatory for over a year, the mere presence of the Astronomer Royal caused him to quake. He trembled so violently that he dropped a bottle of photographic chemicals used to develop the daily solar photographs. This particular incident impelled Airy to write in complaint of the appointment to the secretary of the Admiralty. Eventually, Maunder adjusted to the regime at Greenwich, supported in his astronomical aspirations by his family, particularly his older brother, Thomas. A couple of years after his appointment, Maunder married Edith Hannah Bustin at the Wesleyan Chapel, Wandsworth, and they started a family.

Around this time, grave news from India put unexpected pressure on the astronomers. The monsoon rains had failed to appear and a famine of biblical proportions was sweeping the mighty subcontinent. Millions were dying, and British rule was struggling to maintain control of an increasingly desperate population. Spurred by the scale of the human catastrophe, and in the woefully misplaced belief that this was their chance to show the Indians how to run their country properly,[2] scientists of all persuasions began to search for a way to predict the weather. The Indian Meteorological Department was established at Pune in 1875. It amalgamated a number of regional offices into a sustained, directed workforce.

Mindful that the auroras were atmospheric phenomena, many renewed their interest in William Herschel's turn-of-the-century ideas about a sunspot-climate link. But in trying to prove such a link they came across an immediate problem. What quantity should they measure to characterize the weather: pressure, temperature, rainfall, all of these and more? As data flowed in from the various weather stations that had been set up by John Herschel as the complementary strand of Sabine's magnetic crusade, legions of poorly paid "ciphers," often arithmetically talented children who worked twelve-hour shifts, scrutinized the numbers for anything that appeared to correlate with the sunspot cycle.

Initially, a number of promising coincidences were found. The average air pressure over India seemed to drop to a minimum during the years of sunspot maximum. At the same time, storms in the Indian Ocean became more frequent. More correlations were noticed from other regions of the globe. The sunspot maximum seemed to line up with minimum temperatures being recorded in Scotland and South Africa. In America, sunspot maximum seemed to have triggered large quantities of rainfall

[2] Modern historians trace the origin of these catastrophic Indian famines not solely to the failure of the weather but also to the land seizures by the British. By changing local subsistence farming into plantations for foreign exports, and restricting internal trade, the imperialists siphoned valuable food out of the country and also prevented regions of plenty supplying those most affected by the uneven rainfall of the time. It is said that during the worst of the famines, there was food available in the markets but that the poor could not afford the inflated prices.

with the result that the Great Lakes were swollen to an all-time high. So strong did the suspicions of a link become that, by the end of the decade, children were being taught in school that the weather was modulated by the appearance of sunspots. But the details of the process remained stubbornly elusive.

Mathematician William Stanley Jevons saw the sunspot cycle reaching through the climate and into the economy. Since the middle of the nineteenth century, the Victorians had perceived that commerce moves in cycles of approximately ten years. The cause of the ups and downs remained unexplained, and Jevons struck upon the idea that the decadal economic cycle was similar in length to the sunspot eleven-year cycle. Jevons began his investigation by resurrecting William Herschel's line of inquiry into harvests. He looked at historical data for the price of wheat, barley, oats, beans, peas, vetches, and rye. By averaging their price, he was astonished to find that there did indeed seem to be a repeating eleven-year period in the data, with the maximum price of the commodity falling in the third or fourth year of each cycle. Excited by the correlation, he rushed to publish his findings in the newly established scientific journal, *Nature*.

His work was immediately criticized for relying on poorly defined data and woolly mathematical conditions. In his effort to substantiate the result, Jevons too found that the way he chose to average the prices affected the results. If he changed the criteria slightly, he found different periods of regularity. Lacking any conclusive evidence that the sunspots did affect the British harvests directly, Jevons went in search of other connections.

As the British struggled through the economic depression of the 1870s and India suffered its climatic disasters, Jevons began working on a new hypothesis. With the now widespread belief that the sunspot cycle affected climate, he postulated that in years of famine, the demand dropped for English goods in Indian and other tropical countries, precipitating the economic gloom. So, commercial cycles could still be linked to sunspot maxima and minima; just not directly.

His new idea did little to silence his critics, some of whom were astronomers who felt his inferences went too far. Some blunt rejoinders were published in *The Times* and *The Economist*, pointing out inconsistencies in his analysis. Jevons stood up for his theory and continued to refine it throughout the few remaining years of his life. In the year of his death, 1882, he published again on the subject, but his new analysis failed to stir anyone because interest in the putative links between sunspots and climate had taken a mighty blow just the year before.

The government's chief Indian meteorologist, H. F. Blandford, reported to the Famine Commission that a simple correlation between sunspot numbers and the extent of the monsoon did not exist. Despite the fact that patterns could be made in retrospect, any weather predictions based on solar observations swiftly became inaccurate. If there was a link, according to Blandford, it was not obvious, direct, or useful. Shortly afterward, he proved that the amount of snow on the Himalayas could be used to predict the rainfall of the monsoon, making the world's first long-range weather forecast in the process. The success of Blandford's method almost entirely diverted attention away from the sunspots as a weather predictor.

Bowing out at the same time as interest in the sunspot-climate link was the ever-skeptical Airy. After forty-six years as Astronomer Royal, the eighty-year-old retired in 1881. He moved from the official residence at the observatory just down the hill to the outskirts of Greenwich Park, from where he could still keep an eye on proceedings. He began working toward completing his lunar theory, in which he hoped to explain the measured perturbations of the Moon's orbit by calculating the gravitational influence from all the other celestial bodies in the Solar System.

By this time, Maunder had come to respect Airy professionally but regarded his methods as despotic, and was doubtlessly relieved by the man's departure. Maunder continued his daily photographs whenever the British climate cooperated and found an ally in the new Astronomer Royal, William Christie. Despite having been Airy's chief assistant, Christie was genuinely surprised to be chosen as successor. He was a mild-mannered man of clear perception and was instructed by the Greenwich board of visitors to quietly begin a program of modernization. He almost immediately elevated solar astronomy higher up their list of priorities, ordering new equipment to allow Maunder to measure the area of sunspots and altering the photoheliograph so that Maunder could more successfully track day-to-day changes in the sunspots.

Now head of the solar department at Greenwich, Maunder made his way into the solar observatory twice every clement day and cranked the

E. Walter Maunder, the intellectual heir to Richard Carrington. (Image: Royal Astronomical Society)

steeple-roof around to face the Sun. The photoheliograph telescope had an aperture of four inches, which Maunder often baffled down to just three inches on sunny days. To further cut down the sunlight and prevent the photographic plate from becoming overexposed, the photoheliograph was designed with a clever device that could achieve exposure times of just one-thousandth of a second. A brass plate containing a narrow slit sat in a groove running across the diameter of the telescope. To arm the telescope to take a photograph, Maunder locked the brass plate into position so that it entirely blocked the sunlight from entering the camera portion at the far end. Once he had pointed the telescope at the Sun and securely loaded the film plate, he released the catch. A powerful spring propelled the brass plate downward, sweeping the slit across the focused beam of light, just six-tenths of an inch in diameter, allowing a tiny fraction of light to pass through. A lens then magnified the light to produce an image of the solar surface fully eight inches in diameter. This fell onto the film plate, which Maunder then spirited away to develop.

Maunder became enthralled by the endless machinations of the solar surface, stating that the striking alterations he witnessed provided a source of unfailing interest. He came to think romantically of this task as capturing the Sun's portrait, while the tremulous magnets, in the darkened cellar of the Greenwich magnetic pavilion, recorded the Sun's autograph.

As the years passed, Maunder grew in both knowledge and confidence, transforming from the tongue-tied young man of Airy's day into a man with "a suavity of manner and softness of speech" that those who knew him felt was entirely "in keeping with a kindness of heart and amiability of character that lay deeper than its outward manifestation." He also developed considerable powers of erudition and began to write for publications that communicated astronomy to a wider audience. He also presented public lectures with the same goal in mind.

In a morning mist of November 1882, Maunder saw a sight that transported him back to his childhood. He looked out from his offices at Greenwich to see the Sun shining a dull red behind the mist and a gargantuan spot squatting on its surface. Maunder's gaze shifted to the soldiers marching across Blackheath Common on their way to a great parade in Hyde Park, near Buckingham Palace, to honor Queen Victoria. As they marched, the soldiers pointed out the sunspot to one another. Mindful of the spectroscopic flares that astronomers often reported seeing, Maunder hastened to the large dome where the twenty-foot-long spectroscopic telescope lay at rest during the day. Maunder roused the great instrument and wrestled it into position. He once wrote that taking a spectrum of a sunspot was like seeing into its soul as each spot revealed some unique aspect of itself. Sure enough, as he focused the spectroscope directly onto the solar bruise, he saw one of its hidden secrets. Brilliant tendrils of hydrogen gas were being forced from the spot's vicinity, as if expelled under great pressure—a spectroscopic flare was taking place before his eyes. That night, a powerful aurora lit the frosty winter sky and the telegraph network collapsed. Checking the magnetic readings the next morning, Maunder found that the Greenwich instruments had spent an uneasy night, greatly disturbed by the mysterious magnetism that seemed to be coming from the sunspots. He began to wonder how he could transform these coincidences into some form of mathematical certainty.

As he looked back over the decades of available data, he saw that the general trend was unmistakable, but, whenever anyone tried to dig into the details of the process, the connection always failed. Not every large sunspot produced a magnetic storm, while at other times even modest spots precipitated large disturbances of the magnetic equipment. The key had to be the flares. Only when a sunspot flared did it somehow project its magnetism to Earth, with large spots seemingly more likely to flare than smaller ones. But the biggest puzzle was why the magnetic storms took place nearly a day after the flare. Could the Sun be releasing a cannon ball of magnetism that lumbered across space before striking the Earth? Such a concept was science fiction to the Victorians, who thought that magnetism only surrounded a magnet and so could not be "bottled" and transported somewhere without moving the magnet itself.

Christie had arranged for various meteorological stations across the British Empire to be equipped with photohelioscopes. Their photographs were shipped back to Greenwich, ensuring an unbroken record of the Sun's face. Yet, without the insight into how to mathematically analyze these observations, Maunder seemed doomed to be little more than the Sun's librarian, counting in and out its sunspots on a daily basis, but never truly understanding what he was seeing. Nearly three decades after Sabine's first announcement of the link, and despite the mountain of data at Maunder's disposal, the cause of the magnetic connection between the Sun and the Earth was as intractable as ever. Undoubtedly, Maunder's lack of formal mathematical education was hindering him in analyzing the data but so too was his workload at Greenwich. As well as his solar studies, he was expected to take spectroscopic readings of the stars at night. Beyond the practice of astronomy, his literary skills had made him the obvious choice to edit the Greenwich astronomy magazine, *The Observatory*.

For all the early promise the job had offered, Maunder was beginning to feel trapped. Hearing of the progress made by amateurs and continental professionals in the realm of astrophysics compounded his frustration. In the stars, they had found both similar and different Fraun-

The Royal Observatory at Greenwich. (Image: Royal Astronomical Society)

hofer lines to the Sun, allowing them to begin placing stars into different classifications. They had proven that some of the ghostly nebulas were tenuous clouds of gas, and that some contained new stars, leading them to confirm William Herschel's speculation that the nebulas were the stellar nurseries of the cosmos. In solar physics too, they had identified chemical element after chemical element, many of them metals on Earth. On the Sun these metals were detected in their gaseous form, and some now thought of the Sun's atmosphere as a mirror of the Earth's but with clouds of metallic vapors. When it rained from these clouds, deluges of molten metal fell from the Promethean solar skies.

His endless string of duties left Maunder with no time to stop and ponder where his own research should go and how he could best contribute to this effort. He was the country's first professional astrophysicist, yet he was strait-jacketed by the requirements of his station. While he admired the work of the various Fellows in the RAS, he found himself at loggerheads with the ruling council. Ever since his invitation to become a Fellow in 1875, he had lobbied for women to be allowed to join the society's ranks. Under special invitation they could attend the society's meetings, but it remained unconscionable that they should be allowed to actually join. Despite many attempts to raise this subject, Maunder failed to make headway.

His professional frustration was joined by real tragedy in 1888, when Edith died of tuberculosis, leaving Maunder with five surviving children to care for. His grief fed his growing depression, and soon he came to think of himself as having nothing to offer science. Others disagreed and urged Maunder to divert his enviable talent in communicating astronomy toward founding an organization that promoted the subject to anyone who wanted to learn. The idea resonated with Maunder's deepest sense of equitability and, with his loyal brother's considerable help, he founded the British Astronomical Association (BAA) in 1890. Entry was open to anyone, male or female, who had an interest in astronomy, from the casual to the committed observer. Part of Maunder's motivation was the increasingly technical and mathematical nature of the work practiced by the Fellows of the RAS. He specifically cited that the BAA would cater to those who found the papers of the RAS "too advanced."

Around this time, his reputation as an excellent observer reached America, and he received an invitation to travel to California. The offer must have seemed like a dream come true; but when Maunder requested the time off, Christie refused him the opportunity because there was no one carry on his work at the observatory. This returned Maunder to despair, and he wrote to the director of Lick Observatory declining the offer, adding that Greenwich "has fallen on evil days, and I am not the only Assistant who would be glad to have a fuller opportunity than it affords of doing good work for Science."

Perhaps as a consequence of making these feelings public, and after fifteen years of struggling on his own at Greenwich, Maunder was finally assigned a staff member to help him with the mathematical demands of his job. Christie wanted to emulate the success of an American innovation: the lady computer, in which highly educated, mathematically literate young ladies would perform calculations under the direction of the men. The pay was miserly and the work relatively uninspiring, but it represented the first time women had been allowed into the employ of the Royal Observatory. Annie Scott Dill Russell took full advantage of this opportunity.

She arrived newly graduated in mathematics from Girton College,

Cambridge, and was placed under the tutelage of Maunder. His beliefs of male and female equality meant that he instantly viewed her as a working partner rather than a drudge, and it is certain that he spent many hours on her tuition. She repaid this attention by developing a keen interest in Maunder's solar work, and together they became a symbiotic partnership combining Maunder's fifteen years of observational and astronomical experience with Russell's mathematical skills.

The solar photograph of 15 November 1891 revealed to Maunder a large spot edging around the Sun's eastern edge. In the coming days, two other, separate spots appeared as well. As the trio slid across the Sun's face, Maunder and Russell watched them multiply into groups, dividing like living cells under a biologist's microscope. They marveled at the mighty forces that must be at work beneath the Sun's surface and reveled in the spots' complex beauty until they slid, out of sight, onto the far side of the Sun.

Maunder knew this was probably not the last he had seen of these particular sunspots. Biding his time, his patience was rewarded on 12 December when the Sun's rotation brought one of the groups back into view. Of the other two clusters, one had transformed itself from dark sunspots into a clump of bright patches, while the other had disappeared altogether. A day later, the missing cluster spontaneously reformed, exactly where Maunder calculated it had been the month before. Growing with extreme rapidity, it split into two just before rotating out of view again onto the far side of the Sun.

Battling the winter weather, Maunder and Russell glimpsed one group's third passage in January and managed to track it during just six days on its next return in February. By now it had grown substantially, become a yawning cavity that much reminded Maunder of the 1882 spots. Digging through his records, Maunder realized that it outsized even those behemoths and was the largest spot ever photographed from Greenwich.

As it edged across the Sun, Maunder wondered whether it would unleash an auroral storm. He was not disappointed. The day before Valentine's, the sky beat with a red pulse as the spot rained down its magnetic power. Telegraph operators suffered another day of disruption, and the recently invented telephone lines were rendered inoperable by harsh clicking and ringing tones on the lines. In Princeton, New

Jersey, townsfolk and students gathered on the streets to watch the spectacle, some proclaiming that a mighty calamity had befallen the world.

Maunder was not the only scientist watching the February 1892 spot in rapt fascination. In America, an impetuous young man was risking his marriage in pursuit of the Sun's secrets.

ELEVEN

New Flare, New Storm, New Understanding, 1892–1909

George Ellery Hale had married Evelina, his teenage sweetheart, two days after graduating from the Massachusetts Institute of Technology. He had taken her to live with his parents in the fashionable Chicago suburb of Kenwood, the family having grown rich from selling elevators to builders in the wake of the devastating 1871 Chicago fire.

Despite his academic credentials being only newly minted, Hale was already recognized as a world-class astronomer. Before graduating, he had thoroughly investigated Bunsen and Kirchhoff's technique of spectral analysis and used the concept to invent a new instrument, the spectroheliosope. This was a step up from the Kew photoheliograph in that it could take pictures of the Sun in a single wavelength of light, significantly reducing the glare to show rich patterns of detail. As word of the invention spread, offers of employment from universities across America and Europe came Hale's way. Wary of being used by academics who wanted his knowledge rather than him, he discussed the offers with his father, who eventually decided that the best way to protect his son from exploitation would be to furnish him with everything he needed to set up an observatory on the grounds of their Kenwood home.

Building work commenced on a three-story dome and office complex and culminated in a dedication ceremony in 1891, attended by over one hundred guests, many of them distinguished American astronomers and academics. Assisted by his enchanted siblings, Martha and William, Hale began an exhausting program of solar study that left Evelina feeling excluded. Even her relationship with Hale's mother was strained. The matriarch suffered from migraines and insisted that the family

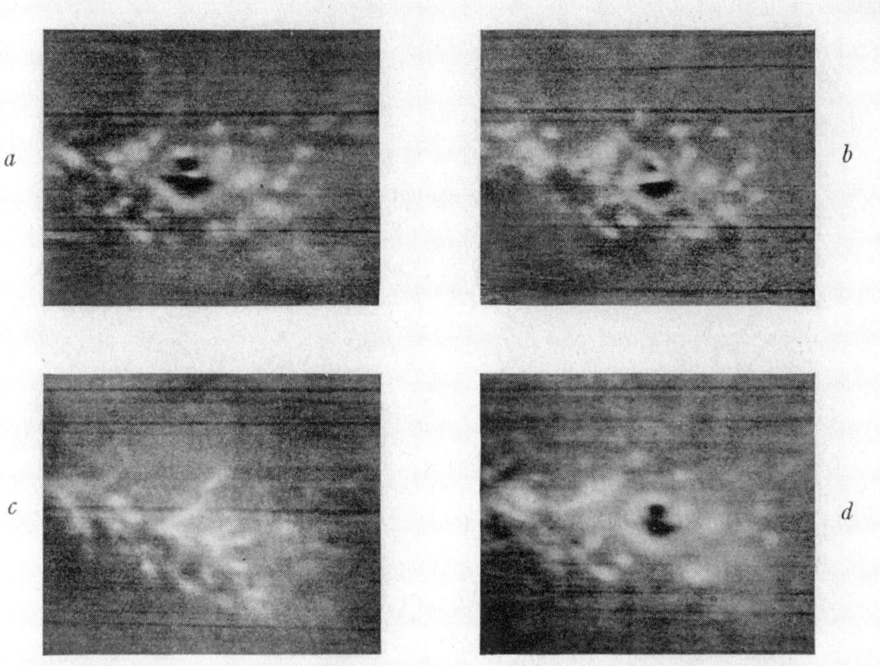

The solar flare of July 1892, photographed by George Ellery Hale from his private Kenwood observatory. The sequence shows the sunspot group before, during, and after the flare. The sunspot group is the collection of dark spots in the center of image (a), taken at 4:58 P.M. The flare itself is the bright bar extending across the dark sunspots in image (b), taken at 5:10 P.M. The aftermath of the flare set the whole area above the sunspot glowing, as seen in image (c), taken at 5:37 P.M. By 7:50 P.M., the area had returned to normal, as seen in image (d). (Image: Hale, George E. (1931) The spectroheliosope and its work, *Astrophysical Journal* 73: 239, Plate IV. Reproduced by permission of the AAS)

house be kept darkened and quiet, so Evelina's only refuge was to sit in the observatory and observe her husband's industry.

The spots of February 1892 provided Hale with an excellent opportunity to perfect his single-wavelength photographs of the solar surface. By the time another complex sunspot group appeared during July, he was routinely taking and developing between five and ten photographs a day. He began photographing one particular sunspot on 8 July when it rounded the Sun's eastern edge. As the days went by, the spot divided into two, and on 15 July, as it neared the Sun's central meridian, a bright ridge of glowing gas appeared between the two distinct spots. Hale's first photograph of that day was taken in the afternoon, his second just twelve minutes later. He made some adjustments to the telescope and exposed a third photograph twenty-seven minutes after that. As he developed the images he saw the spots as he expected to see them on the first image, but the second showed something altogether different. Above the pair was a brilliant streak of light, extending outward into space and terminating in a burning sphere of white. Astounded, Hale developed his third photograph. The bright streak had disappeared, but the sunspots were now completely engulfed by a glowing curtain of gas.

He hurried back to the telescope and rigged it for visual use. Although he saw glowing hydrogen clouds surrounding the sunspot gently radiating the last of their explosive energy into space, like the embers of a dying fire, the scene had virtually returned to normal. It was a near repeat of Carrington's fateful observation as Hale had witnessed a massive solar flare ripping into space above a sunspot group. His more sensitive equipment had allowed him to track the after-effects of the explosion for longer than Carrington, but the parallels were obvious. A day later there was severe disruption across the communications lines, with 210 volts building up on the lines between New York and Elizabeth, New Jersey.

Being in the right place at the right time, Hale had succeeded in capturing the flare on a photograph with his spectroheliscope, securing his fame in astronomical circles and cementing his lifelong interest in the Sun. Shortly after this success, the newly founded University of Chicago came headhunting. Hale's father, acting as manager, negotiated generous terms. Hale would join the university as a professor of astrophysics, the first time the term had been officially used. His Kenwood observatory would be dismantled and rebuilt on the university's campus, on a site where Hale remembered picking wild strawberries as a boy. In exchange for his relocation, the university would subsequently release funds of not less than $250,000 for Hale to build a second and better observatory.

Evelina welcomed the move because it would extricate her from the suffocating grip of Hale's family home and allow her to establish a social life with the wives of other academics at the university.

The new observatory was subsequently built at Williams Bay, Wisconsin, and named after the disgraced banker Charles Yerkes, who funded the project in the hope that such philanthropy would restore his social standing, having served seven months in prison for the misappropriation of funds. Hale soon founded the American Astronomical Society and the first professional journal dedicated to publishing the results of the new astronomy. It was called, appropriately, *The Astrophysical Journal* and remains a leading journal today.

With Hale's photograph of the flare circulating among British astronomers, too, the endgame had begun. The magnetic Sun-Earth connection erupted onto the scientific agenda of a new generation of astronomers, the original sparring partners having by now joined Carrington and Herschel in their respective graves.

After an ignominious ousting from the Royal Society, Colonel Edward Sabine had brought his series of magnetic catalogs to an end and hung on until he was ninety-three, finally dying in East Sheen during 1883. The former director of the Kew Observatory, Balfour Stewart, had been seriously injured in a railway accident in 1870 but recovered to hold the chair in physics at Manchester until his death in 1887. His great contribution was to expand on the ideas sparked by the magnetic disturbance associated with Carrington's flare and to reason that Earth's upper atmosphere contained an electrically charged shell of gases, known today as the ionosphere. Following the handover of the Kew photoheliograph to Greenwich, and the disbursement of his other telescopes to the Radcliffe Observatory, Oxford, Warren De la Rue had lived in quiet retirement until his end in 1889.

Even the indomitable George Airy had released his grip on the world. He never did complete his lunar theory. When he discovered an error in his early analysis, the thought of reworking all of his mathematics robbed him of the will to continue the work. In his retirement writings, in which he referred to himself in the plural, Airy passed candid judgment on his contribution to science. He wrote, "Our principal progress has been made in the lower branches of astronomy while to the higher branches of science we have not added anything." He passed away in January 1892, just six months before Hale's flare rekindled interest in the link that Airy always rejected.

At the forefront of the new investigators was Maunder. Inspired by the unsurprising appearance of magnetic storms when the February sunspot had crossed the central meridian of the Sun, Maunder set about finally proving the link between magnetic storms and large sunspots. The continuing failure of such investigations was playing into the hands of a new and powerful enemy: Sir William Thomson, otherwise known as Baron Kelvin of Largs.

Lord Kelvin was a scientific colossus. The first scientist elevated to the peerage, he once delivered a lecture at the Institute of Civil Engineers in which he had said, "When you can measure what you are speaking about and express it in numbers, you know something about it." Such a belief had served him well, especially when he used mathematics to help mastermind the laying of the workable transatlantic telegraph cable in 1866, an endeavor that even George Airy had regarded as impossible.

Kelvin could deploy his mathematics as if it were a rapier, cutting to the very heart of a problem. As interest again grew in sunspots giving rise to magnetic storms, Kelvin decided to bring an end—once and for all—to what he believed was scientific mumbo-jumbo. He chose to launch his assault from the highest perch in science: the presidential address to the Royal Society.

Despite the sprouting of new scientific organizations during the fertile springtime of modern science during the nineteenth century, the Royal Society had never lost its preeminent cachet. It was still the one that all scientists aspired to join. Lord Kelvin became its president in 1890, at the age of sixty-six. With his receding snowy hairline and bushy beard, he exuded aged wisdom and spoke with the conviction of utter confidence.

On 30 November 1892, he rose to give his second presidential address. The gathered Fellows and their guests listened as he told them that he hoped to correct fifty years of outstanding difficulty in understanding the supposed connection between the Sun's surface and the magnetic storms on the Earth. He laid the blame for the misconception at the feet of his "ancestor in the Presidential Chair," Edward Sabine, and implied that those in positions of scientific power had misled the scientific community. As evidence, he read from the presidential speech of Lord Armstrong, given before the rival British Association for the Advancement of Science in 1863. The excerpt noted the magnetic spike associated with Carrington's flare, and the ferocity of the subsequent magnetic storms. It suggested that the flare was the collision of a large meteorite with the Sun and that this event had supplied the energy that had radiated itself across space to trigger the magnetic storm on Earth.

It was exactly the kind of woolly reasoning that Kelvin abhorred, and he talked his audience through a strict mathematical analysis of the problem. He explained that since the time of Carrington's flare and the first discussions of a link to magnetic storms, the Scottish theoretician James Clerk Maxwell had developed a succinct quartet of mathematical laws that described the inextricable links between electricity and magnetism. Presenting magnetic data from various observatories, Kelvin made it clear that magnetic storms often exceeded the strength of the Earth's natural magnetism by many times. He then set about calculating how much energy the Sun would need to expend in order to exert this influence across 93 million miles of space. According to Maxwell's laws, the greatest magnetic catastrophe possible was if the Sun's north magnetic pole suddenly became it south magnetic pole, thus reversing the magnetic field. This would send a magnetic shockwave bursting through space in all directions at the speed of light. So the energy recorded in a magnetic storm was just a tiny fraction of the true quantity released into space.

Kelvin calculated that to drive even a moderate magnetic storm on Earth required the Sun to release in a few hours as much energy as that radiated into space during four months of its normal shining. The thought that the Sun could do this and yet remain unchanged in appearance, apart from the occasional dark spot, he considered absurd.

He stated that Carrington's flare was probably nothing more exotic than a fountain of hot solar material jetting upward before falling back to the surface. With this comment Kelvin betrayed that he had not read Carrington's exacting description of the flare but relied only on the brief comments given by the former president of the British Association for the Advancement of Science. Carrington had described how the phenomenon was not connected to the surface of the Sun because the surface remained unchanged even though the flare had skated across 35,000 miles (nearly four and a half times the Earth's diameter) of it in just five minutes. This simple observation proved that the flare had been confined to the atmosphere of the Sun alone.

Oblivious to his mistake, Kelvin urged the audience to forget sunspots and work harder in finding the true connection between the auroras, the magnetic storms, and the Earth currents that simultaneously flowed along the telegraph lines. The comments were published in full in the journal *Nature* and widely distributed.

Although Lord Kelvin's mathematics were difficult to contest, Maunder and others also knew that the statistics linking general sunspot activity to the frequency of magnetic storms was equally irrefutable. Only in the day-to-day detail did the statistical connection break down. There had to be a missing piece to the puzzle. In a gesture of appeasement to the great physicist, some astronomers began to suggest that an unknown "third-party" astronomical phenomenon was affecting both the Sun and Earth, sometimes simultaneously, sometimes individually. On Earth, the influence caused auroras, magnetic storms, and Earth currents, while on the Sun it caused sunspots. Maunder rejected this as unnecessarily complicated and remained convinced that the root cause lay in the Sun. Unfortunately, neither he nor his mathematically talented lady computer, Annie Russell, could see a way of proving their conviction in the language of mathematics that Kelvin would understand. Maunder remained silent save for his daily logging of sunspots and their characteristics.

As he worked day-in, day-out with Russell, Maunder's frustration at having to remain at Greenwich eased, and he began to see that there might be significant compensations. Russell was both a tireless supporter of Maunder's work and an accomplished astronomer in her own right. Beyond her solar work, she used the mathematical laws of optics to design a tiny wide-angle camera. Despite having only a one-and-a-half-inch diameter lens, her calculations showed that it would be capable of recording the faint band of stars that constituted the Milky Way. Her former college at Cambridge awarded the necessary funds to have the camera built.

Maunder enjoyed her successes as much as his own. His Wesleyan upbringing meant that, despite her being both a woman and more than a decade younger than he, he treated her as an equal and appreciated her intellectual skills. Together, they discussed their joint belief in the

Bible and the equality of the sexes. Eventually, Annie Russell became Mrs. Maunder, on 28 December 1895, and joined the family home in Greenwich. Maunder's existing sons and daughters accepted her as their stepmother, and Maunder even referred to her as their mother. For her part, Annie wrote of them affectionately in her letters but never bore a child of her own.

The personal connection brought about by the marriage strengthened their professional relationship and the routine preparation of their sunspot catalogs honed their teamwork. For her, Maunder was her passage into the London scientific societies, most of which still excluded women. For him, Annie was both his muse and his mathematics.

As 1898 approached, Maunder's British Astronomical Association took the lead in organizing a trip to India to view the total solar eclipse that would take place there on 22 January 1898. In contrast to Airy's 1860 expedition, there was no vetting of scientific intentions. This trip was open to anyone who could pay his or her own way. A similar BAA expedition had taken place in 1896 to the Norwegian eclipse. Despite the weather being cloudy, the experience had strengthened the BAA's sense of its own usefulness and brought it a fair amount of public credibility. This time, the Maunders realized that Annie's camera would be the ideal instrument to capture the faint details of the Sun's outer atmosphere. They set this as their primary goal for the upcoming eclipse.

The eclipse track passed through no convenient large town and Maunder therefore decided to head for the village of Masur because it was served by a railway. Shortly before departure, the reports of a virulent outbreak of plague in the region scuppered these plans. Committed to making the journey, Maunder, his wife, and three fellow eclipse chasers boarded the P&O mail steamer RMS Ballaarat at Tilbury on 8 December 1897, with no idea about what would be waiting for them in India.

Onboard ship, the astronomers passed the time by making daily estimations of their longitude and scrutinizing a group of sunspots that edged across the Sun's disk. To accomplish this easily, they positioned themselves so that smoke from the ship's belching funnel crossed the Sun, reducing its glare. In the evenings, as the Sun sank from the sky, they competed with one another to be the first to see the pinpoint glow of Mercury in the twilight. Then they watched for both the diffuse glow of the zodiacal light, caused by sunlight reflecting from dust clouds in space near Earth, and the combined light of distant stars that showed up as the Milky Way. Night after night they strolled the decks, noting the inexorable sinking of the Pole Star behind them and the revelation of unfamiliar southern constellations in front. One member of the group fixed his spectroscope to the underside of his cabin mate's bunk, so that he could test the instrument on the sunlight streaming through the porthole.

They made landfall in India on Monday morning, 3 January 1898, and received word that they were to make their way by rail to the village of Talni, where a camp would be available. The journey began at night to avoid the heat, took eighteen hours, and conveyed them into the dusty plains of central India. Once at the camp, the astronomers began their preparations while maintaining a watch for wild animals. Despite the stories they had been told of restless tigers, panthers, and hamadryads, they saw nothing save for a small snake. Even the nighttime calls that drifted through the camp were explained as nothing more exotic than jackals. For Maunder, the lack of wildlife was something of a disappointment.

At night, Annie set up her wide-angle camera and took pictures of the Milky Way. She developed them in the expedition's darkroom, housed in a wattle and daub mud hut with walls that were "bowed and bent in all kinds of fantastic curves." On eclipse day itself, Annie stood poised at the camera while the men fussed with their own cameras and instrumentation. The temperature dropped, the colors faded from the landscape, and darkness fell. They had just two minutes to get their work done. Immediately Maunder heard a faint wailing coming from those gathered in the village. Over this, the party's timekeeper called out the remaining seconds of totality at ten-second intervals. Jolted into action, Maunder and Annie saw that the corona was bright, so bright that it provided more light than even a full Moon. It was active too, with rays of light reaching out into space. Annie set her camera to take a picture. When light returned, a shout of thanksgiving issued from the villagers. The visiting astronomers felt a great relief, because it had been clear and their experiments had worked.

That evening, Talni erupted in celebration, the villagers giving themselves up to "unrestrained rejoicings," according to Maunder. The eclipse party joined the celebrations and were garlanded with flowers

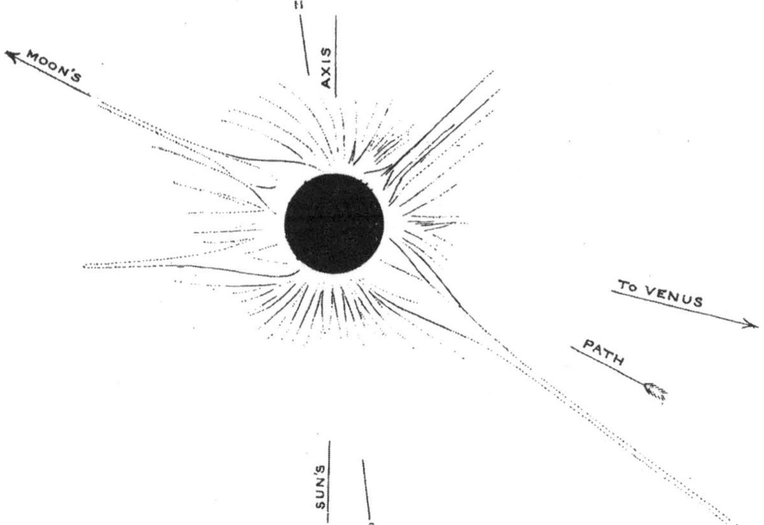

Annie Maunder's photograph of the 1898 eclipse, showing streamers reaching out into space. An accompanying sketch highlights these fainter features. (Image: Stuart Clark, private collection)

and anointed with betel and scent, but Maunder found the native music a little raucous for his taste.

When Annie developed her picture, the husband and wife team knew that the journey had been worth it. She had captured the extraordinary corona, including the straight beams of milky light that extended for many times the Sun's diameter into space. With pencil and paper, Annie deduced that the longest of the streamers reached over 6 million miles into the void. The calculation ignited Maunder's imagination.

He began to wonder whether the beams led all the way back to sunspots. Perhaps his wife had captured a view of the solar rays responsible for the magnetic storms. He imagined the rotation of the Sun sweeping these beams through space like a giant celestial lighthouse. When one happened to be directed toward the Earth, whatever strange emission it contained pummeled the planet, creating the magnetic storms. If true, it would refute Kelvin's assumption that the Sun radiated its magnetic energy uniformly throughout space. By somehow focusing its electromagnetic force into beams, the Sun did not waste any energy, and Kelvin's objection about the outrageous power required was untenable; either the beam hit the Earth, sparking a storm, or it missed. But Maunder knew of no mathematical theory for magnetic behavior of this kind. Without it, his interpretation of the photographs was merely speculation.

Unbeknownst to Maunder, in laboratories at Cambridge, amid the glowing vacuum tubes and humming electricity generators, physicists were making some remarkable progress that would eventually come to his aid.

They were beginning to understand cathode rays. These mysterious rays carried an electrical charge and passed in straight lines through glass tubes that had been pumped free of air. After a series of experiments that showed the cathode rays could be bent by the application of magnetic or electrical forces, Cambridge physicist Joseph John Thomson deduced that the mysterious energy beams were composed of great numbers of negatively charged particles. He called the particles *electrons.* He measured their speed and found that they traveled much more slowly than light. As for the nature of the electrons, Thomson asked, "What are these particles? Are they atoms, or molecules, or matter in a still finer state of subdivision?"

Working at Cambridge at the time was an Irish mathematician called Joseph Larmor. He was the motivating figure in a scientific coup d'état that was gradually gaining momentum among the cathode ray tubes. Larmor was working on a whole new way to think about electricity. He postulated that, instead of Maxwell's ideas of fields radiating electrical energy like ripples on a pond, charged particles carried electricity like the flow of water along a river. With Thomson's discovery of the electron, Larmor's concept began to be accepted. Although there was much work to be done on its refinement, and the full implications of this shift in thinking had yet to be revealed, physicists began to look at the Universe in an entirely different way.

Seemingly oblivious to this new window on the world, the aging Lord Kelvin addressed the British Association for the Advancement of Science as the nineteenth century crossed into the twentieth. He could not resist launching another volley against the sunspot origin of magnetic storms when he sweepingly dismissed any thoughts that there was some unanticipated piece of science that would render the connection explicable. "There is nothing new to be discovered in physics now. All that remains is more and more precise measurement," he foolishly proclaimed.

It took Maunder several more years to experience the epiphany he needed to counter the great weight of Kelvin's opinion. It began in October 1903 when another sunspot cycle reached its climax. A huge spot crossed the face of the Sun but produced only a moderate storm as it passed the center. A fortnight later, a smaller sunspot drew into position and unleashed the largest magnetic storm in the Greenwich records, ensuring that the world's telegraphers suffered yet another miserable day of frustration.

Intrigued by the discrepancy in the size of the spot and its associated magnetic storm, Maunder began digging through the Greenwich records looking for the biggest magnetic disturbances. He found that nineteen great storms had taken place during the preceding thirty years. He then checked them against the sunspot record. For every storm he found a great sunspot lying near the center of the Sun's disk, or a lesser spot that had once been enormous occupying the same place. He then reversed the process, looking for the nineteen largest sunspots during those three decades, back to 1873, and checking the magnetic records against them. This time he found that the nineteen greatest sunspots had produced seven severe magnetic storms, seven considerable ones, two minor storms, two even smaller ones, and one without any disturbance at all.

This told him that magnetic storms needed sunspots, but that the size of a spot could not be used to predict the intensity of any resulting storm. Such behavior was understandable if magnetic disturbances erupted at random from sunspots and were directed along definite paths rather than radiated equally in all directions. If the sunspot ejected its magnetic energy only along certain paths into space, then the severity of the magnetic storm was dictated by other factors. For example, perhaps the path did not quite point at Earth. In that case, even an enormous eruption over a giant sunspot could miss Earth altogether or just sideswipe it, producing nothing but a small-scale disruption of the magnetic compass.

To bolster their conclusion, Maunder and Annie began a comprehensive investigation that included all magnetic storms regardless of size. Their task was made easier because the superintendent of the Magnetical and Meteorological Department at Greenwich, William Ellis, had already classified the magnetic storms into "Great," "Active," "Moderate," and "Minor" categories. Between the years 1848 to 1881, the Greenwich magnets had recorded 276 storms.

For eight months, Maunder and Annie laboriously checked the sunspot records for each day that a storm of any size had taken place. Although the results looked promising for the great storms, it was impossible to extricate similar correlations for the smaller storms. Some days they happened when no spots were visible, other days when there were multiple spots, leading to the problem of which spot to associate with the storm. So, while the large spots were clearly associated with storms, without a way to chase that connection down to the smaller scales there seemed no means for the Maunders to prove that all the magnetic storms emanated from the Sun.

Then he saw it.

Near the end of 1886, four successive storms had struck the Earth, each one separated by the same interval: twenty-seven days. Scanning the catalog, he found another series of four consecutive storms in the next year, again each separated by twenty-seven days. That number resonated deeply because it was the average rotation period of the Sun as seen from the Earth. In other words, every twenty-seven days, the same patch of the solar surface faced the Earth. It was the breakthrough

Maunder needed. He realized instantly that you did not need the sunspot data to prove the connection to the Sun; everything you needed was in the magnetic storm data. If he found that storms often followed one another in twenty-seven-day cycles, that would be enough to establish the Sun as their root. Nothing else in the Universe fell into step with the Earth in that particular time period.

A twenty-seven-day cycle would also disprove the idea that the Sun radiated the magnetism from its entire surface because, in that case, there would be no tendency for the storms to follow the Sun's rotation period. The twenty-seven-day periodicity spoke clearly that magnetic energy was being released in directed beams from specific areas of the Sun's surface.

Back in the 1850s, Richard Carrington had developed an equation to calculate the Sun's longitude at any given time. Maunder looked up the formula and, together with Annie, began calculating the longitude of the Sun facing Earth at the commencement of each magnetic storm recorded by Greenwich. They found that once a storm had erupted from a particular solar longitude, there was a strong tendency for it to do so again twenty-seven days later. Of the 279 storms analyzed, one-third came in such pairs. In eight of those cases, a third storm followed on the next solar rotation, and four of those went on to produce a fourth. In one spectacular case, six successive eruptions followed the twenty-seven-day pattern. Still more joined the pattern when the same longitude returned, not in the next rotation but the one after that, giving a fifty-four-day recurrence period.

This was exactly the mathematical proof Maunder needed to refute Lord Kelvin, and he began to prepare the calculations for presentation at both the Royal Astronomical Society and the British Astronomical Association. As word of his breakthrough spread, many grew curious to hear this direct challenge to Lord Kelvin. On the afternoon of Friday, 11 November 1904, the waiting came to an end. The Fellows of the Royal Astronomical Society flocked to Burlington House on London's Piccadilly to hear Maunder present his ideas.

In the confines of the lecture hall, Maunder talked the fellows through his reasoning, clearly making his case that the twenty-seven-day recurrence tied the solar storms to specific regions of the Sun's surface. He then projected his wife's eclipse picture showing the coronal rays stretching through space and referred to the work of the Swedish Nobel laureate, Svante August Arrhenius, who had recently suggested that particles carrying electrical charges could, under certain circumstances, be blown from the Sun, in the manner of comet tails. He wondered if this is what his wife had caught in her 1898 photograph? He then showed another picture, taken by Annie in 1901 from Mauritius during the total eclipse of 18 May. It was a close-up of the Sun's southwestern quadrant and revealed fountains of coronal gas reaching up from the Sun's surface to escape into space.

In the reflected glow of the projector screen, Maunder finished his presentation with a bold claim. "I would suggest that, in the results I have here brought forward, we are on the track of solving what Lord Kelvin spoke of twelve years ago as 'the 50-year' outstanding difficulty."

The president of the RAS, Professor H. H. Turner, thanked Maunder for such an important presentation and, although time was pressing, he opened the floor for a full discussion. First to register his disbelief was Father Aloysius Laurence Cortie, who was associated with the Catholic college at Stonyhurst, Lancashire. He had hurried to the meeting that afternoon when he learned of Maunder's topic. Apologizing that he could not offer a more detailed criticism of the work, Cortie stated that he had heard nothing that convinced him that Lord Kelvin's objection was in error. Others joined in the skepticism by wondering how the details of the process could work. For example, did the Earth attract these rays to it? How far did the rays extend into space? How did the Earth discharge the electricity it received from such rays? Without such answers, they felt uncomfortable placing much stock in Maunder's words.

Still others rose to defend Maunder. Sir Robert Ball told the audience that it should feel under a great debt of gratitude to Maunder for providing the final, irrefutable evidence of the Earth's magnetic links to the Sun, and that the presentation should be one that all in attendance should never forget. Sensing the growing confusion among the Fellows, University of Cambridge astronomer Hugh Frank Newall proposed that when papers of such magnitude were to be presented at the RAS, they should be circulated before, enabling Fellows to arrive having studied the contentious ideas.

Maunder brought the discussion to a close with a final courteous address. He told the Fellows that the paper must stand or fall on the fullness of the analysis and that he realized the meeting could not simply be expected to take his word. He therefore hoped that, when they had the opportunity to study the published paper, they would agree with his interpretation. That proved to be somewhat optimistic.

At the January meeting of 1905, Maunder returned to the RAS to hear his critics. Objections to his paper had crowded virtually everything else off the agenda. In conversation with the Society's president beforehand, he joked that he had hoped the paper would be "severely heckled." Behind this veneer of jocularity, Maunder must have known how serious the meeting was. Maunder was on trial as a scientific heretic. Kelvin's acolytes would put on a fierce show to cow him and his ideas. If he displayed anything but utter confidence in his own work, it would cast doubt on its validity. To help Maunder prepare, the president allowed him to see the chief opponent's paper the night before the meeting. Maunder then spent the next day composing his defense against Professor Arthur Schuster of the Victoria University of Manchester. The professor was not at the meeting in person but had sent his paper so that the president could read it out.

Schuster was a feted pioneer of periodgram analysis, a way of looking for repeating patterns in long strings of data. If the Maunders' analysis was going to be accepted, it was vital that Schuster agreed with their methodology, because a simple periodgram analysis was exactly what they had used to reach their conclusions.[1]

Inside Burlington House, Father Cortie opened the volley with his promised critique, but it consisted of little more than whittling at details in the hopes that Maunder's thesis would somehow topple under this meekest of assaults. The president rose next and began to read from Schuster's paper. After a lengthy discussion of periodgram analysis techniques and analogies, Schuster's main conclusion was that

[1] Schuster had recently presented a number of highly theoretical papers on periodgram analysis to the Royal Society. As an undergraduate, he had begun his scientific career with an apparently more experimental approach. During the height of the late 1870s' discussion of sunspots and climate connections, Schuster had reported that the years of good wine vintage in western Europe occurred in intervals of approximately eleven years. Whether this was a real connection or some undergraduate devilment remains unknown.

Maunder might be right. The professor begrudgingly agreed that one interpretation of the data was a twenty-seven-day period of succession. Nevertheless, he could not bring himself to believe that the magnetic storms originated on the Sun. Lord Kelvin's objections simply ran too deeply for him to accept this, and he chided Maunder for his "somewhat boastful claim" to have solved Kelvin's problem.

Dr. Newall from Cambridge then confessed that he found periodgram analysis too difficult to understand and had therefore spent the day trying to understand the validity of the Maunders' work by arranging colored books in his office. Books with red spines represented days of magnetic storms and everything else days with no storms. By placing them at random in a bookcase, he wondered whether by chance he could reproduce a sequence similar to that which Maunder saw in the magnetic storms. If he could, it would invalidate Maunder's claims. Almost needless to say, Newall ran out of books and shelf space and was forced to attempt to complete the experiment in his imagination. The point he was attempting to make was that he thought Maunder's choice of the Sun's rotation period was too arbitrary and that he wanted Maunder to try analyzing for other periods, too.

The naysayers ended up arguing among themselves, with Father Cortie feeling that he was being misunderstood and others wanting to debate the validity of Schuster's periodgram analysis. In the midst of this confusion, a distinguished visitor to the Society rose to speak. He was none other than Professor Joseph Larmor, the recently appointed Lucasian Professor of Mathematics and a guest of the Greenwich astronomer Frank Dyson. Larmor was still developing his theory of electricity that championed the flow of electrons over the rippling of electromagnetic energy.

He explained that he had not intended to make a contribution to the meeting but, listening to the timbre of the discussion, felt duty bound to speak in Maunder's defense. Larmor had seen the original November paper and spent a whole week working through the logic. He was well acquainted with periodgram analysis and told the gathered Fellows that the statistics of Maunder's association was both convincing and strong. In short, Larmor stated, there must be a connection to the Sun, the only question was the nature of the connection. For an explanation he turned to the burgeoning new theory of the electron and told of how rivers of electrons could equally carry electromagnetism in a single direction. In Maunder's analysis such rivers were clearly implicated, and in Annie's pictures, such rivers appeared to have been captured.

Larmor did offer one suggestion to Maunder with respect to the latter's championing of Arrhenius's particle theory. Over a decade earlier, back in 1892, the Irish physicist George Francis FitzGerald had suggested that sunspots could be the origin of "some emanation like a comet's tail" and that, once ejected into space by the explosive conditions observed to take place over sunspots, this emanation could cross space in a day or so to sometimes strike Earth. It had been a remarkably prescient suggestion but, coming in the same month as Kelvin's heavy publicized criticisms of the field, had gone largely unnoticed. Thanks to Maunder's analysis, and the experimental work at Cambridge, Larmor believed that it was time for this hypothesis to be taken seriously. He was convinced that the smallest particles known to humanity were the agents of interaction between the Sun and the Earth, and with this realization a whole new universe of possibilities was opening before their eyes. Investigations of this rich tapestry of particle interactions beckoned, and would distinguish the twentieth century from the nineteenth.

Lord Kelvin, whose 1892 remarks had sparked the showdown, remained silent on the matter. Now in his dotage, he conceded nothing to Maunder but neither did he contradict him. The naysayers rumbled on throughout the next year. Father Sidgreaves, a colleague of Father Cortie, highlighted two cases from late 1889 when a pair of magnetic disturbances had burst from a spot-free Sun. Where did this leave Maunder's theory? Maunder answered skillfully by showing that this pair of magnetic storms had taken place when a specific solar longitude had drawn opposite the Earth, and on the previous rotations that longitude had played host to a large group of sunspots. Father Sidgreaves's two disturbances had been the parting shots in a great sequence of six storms. Maunder concluded that regions of the Sun could remain magnetically active after the visible spots had subsided. He also found examples of other "blind" eruptions that presaged the appearance of sunspots, indicating the buildup of magnetic activity that, at its peak, caused sunspots to open.

In America, George Hale heard of Maunder's work and felt left behind. Hale's photograph of the 1892 flare had focused astronomers' attention on the Sun, but over the succeeding decade he had become so ensnared with the administration of the Yerkes observatory that he felt he had accomplished nothing in solar research. "New ideas come to me very slowly, and only as the result of continual thinking in and out of working hours," he wrote to a colleague. The day-to-day running of Yerkes did not allow for this. His frustration led him to explore California, where he found a perfect site for an observatory on Mount Wilson.

In 1905, he decided to extricate himself from Yerkes, establish a new observatory dedicated to solar research in California, and make up for lost time. He had personal reasons for moving, too. His eight-year-old daughter Margaret was unwell. He had left Margaret and her mother, Evelina, in California in the hope that the warmer air would help the girl's ailments. His hopes of joining them were dashed when the University of Chicago's president, William Rainy Harper, refused to accept his resignation.

Harper considered that Hale was abandoning ship at the very moment Yerkes needed him the most. Charles Yerkes had withdrawn his financial support for the observatory so that he could move to London and invest in the underground railway that was being developed there. What particularly angered Harper was that Hale had secured $300,000 from the Carnegie Institute to build the Mount Wilson Solar Observatory. It was money that Yerkes badly needed. The affair turned acrimonious and dragged on for months until the university's board of trustees intervened to allow Hale to take the money and move on.

Once in California, the Mount Wilson observatory took shape quickly and Hale set about reestablishing his credentials as a world-class solar observer. His dedication paid off three years later when he made the first detection of strong magnetic fields in sunspots. Then, on 10 September 1908, he saw the first stirrings of activity over a sunspot. Recognizing the behavior from the 1892 flare, he watched carefully and photographed the resulting solar flare. Hale publicized the tell-tale signs to watch for, namely a buildup in the brightness of the clouds surrounding particularly complex sunspots, and other researchers began to recognize the indications of an impending flare. On 12 May 1909, one of Hale's colleagues at Mount Wilson caught a fresh eruption, and later that year, yet another flare was seen from London. In each case, a major magnetic storm took place around a day later. The link was clear.

All the groping for certainty that had begun on the morning of 1 September 1859, when Richard Carrington had seen the unprecedented skittering of light across a giant sunspot, was at an end. The Earth was not an isolated globe in space. It was subject to the whim of the Sun, which could release a strange type of weather in the form of clouds of electrically charged particles that blew through space, igniting magnetic storms and auroras. Now scientists could devote themselves wholly to investigating this most exotic of processes, opening new frontiers in science along the way.

TWELVE

The Waiting Game

William Ellis had begun his scientific career as Carrington's successor at the Durham observatory back in 1852. He ended it as a colleague of Maunder's at Greenwich, supervising the magnetical and meteorology department. As such, he was in charge of logging the day-to-day magnetic records. His only false move had been the day he had visited a nearby power station and returned having unwittingly magnetized his umbrella. For a week, his staff puzzled over the peculiar deflection of the magnetic needles that occurred daily between 9 A.M. and 3 P.M. The truth was swiftly discovered when an alert staff member remarked at the coincidence between the disturbance and Ellis's working hours.

As Maunder was preparing the evidence that magnetic storms were launched from specific sunspots in random directions, Ellis realized one humbling implication. The Earth, large as it may seem from the surface, is actually a small target in space. Most of the Sun's magnetic salvos must miss our planet completely, and therefore were being lost to human eyes. Earthbound observers were effectively peeping through a keyhole and trying to describe the room beyond. In 1904, on the eve of his retirement, Ellis stood before the Royal Astronomical Society and wished out loud: "If we could plant observatories on some of the other planets of our system, and communicate therewith, how enlarged might be our knowledge of the action of the forces that surround us."

Nearly a century later, astronomers effectively achieved this during the Halloween flares of 2003. The magnetic instruments were not in fact sitting on planetary surfaces but voyaging through space on robotic probes. As the Sun blasted its fury in all directions, the separate spacecraft recorded as much as they could.

After the enormous cloud of seething gas swept across Earth on 29 October 2003, it continued onward. One and a half times farther from the Sun than the Earth, it encountered Mars, its rage virtually undiminished. In orbit around the red planet was NASA's *Mars Odyssey* spacecraft. It was mapping the planet and measuring the levels of radiation that astronauts would have to endure on a visit. As the electrified cloud engulfed Mars and its orbiting visitor, the radiation monitor burned out, overloaded by the very phenomenon it was designed to measure. Other instruments on *Mars Odyssey* watched the blast wave contort the tenuous Martian atmosphere to the breaking point and rip off a large chunk to carry with it, into the oblivion of deep space. Amazed scientists realized that Earth's inherent cloak of magnetism had been the only thing that saved our atmosphere from a similar assault.

One of that fortnight's solar eruptions struck the giant planet Jupiter, five times farther away and in quite a different direction than the Earth. It sparked auroras on that planet and a massive magnetic storm that blasted angry radio waves into space for a week. The *Ulysses* spacecraft, a joint European-American craft that was using Jupiter's substantial gravity to lob it back toward the Sun, caught the full force of these radio waves. The minibus-sized *Cassini* spacecraft recorded a similar magnetic event as it approached the beautiful ringed planet of Saturn, ten times farther from the Sun and in another direction again than the Earth.

Leaving the planets behind, the eruptions of gas headed off into the depths of the solar system, gradually spreading out and weakening as they went. Yet, if scientists thought they had heard the last of the storms, they were mistaken. In April 2004, the clouds that had once been a part of the Sun caught up with the aging *Voyager 2* spacecraft. During the 1980s, *Voyager 2* had shown the world its first close-up views of Jupiter, Saturn, Uranus, and Neptune. Now it was 7 billion miles away, unable to return home because of the speed it had acquired during its planetary encounters. The shock wave that washed over *Voyager 2*, although diminished in strength, was still powerful, and astronomers realized that it would have a profound effect when it reached the very edges of the solar system.

Just as the Earth has its cloak of magnetism, so too does the Sun. The Sun's magnetic field extends beyond the planets out to a total distance of some 12 billion miles. The energy carried by the Halloween storms would bolster the Sun's magnetic bubble, expanding it by a further 400 million miles.

The realization presented a staggering new way to think of the Galaxy. Beyond the Sun's magnetic field lies the magnetic influence of the other stars. No longer was deep space a realm of individual stars widely scattered through the void like bright islands in a sea of black. Now it was a place of vast magnetic domains, each centered on a star and pulsing in time to the beat of the star's magnetic heart.

Just as amazingly, the spacecraft that had revealed this new view of the Galaxy relied on technology that had been made possible by the very same revolution that had made magnetic storms understandable. It was the one that had been started by mathematician Joseph Larmor and the particle experimenters at Cambridge. They looked into the world of the very small and found it populated with particles that join together to make atoms. At the center of an atom sits a collection of heavy particles known as protons and neutrons. The protons carry positive electrical charges and the neutrons carry no charge. J. J. Thomson's electrons carry negative charges, are the lightest constituents of atoms, and orbit the nucleus.

The behavior of this submicroscopic world was finally described by a set of mathematical equations developed by a collection of European physicists during the 1920s and was called *quantum theory*. It represented a fundamentally different way to look at physics. Instead of forces being visualized on the large scale as three-dimensional volumes, known as fields, that were capable of inducing straying objects to move, quantum theory described forces as being carried bit by bit on particles. The collision of these particles resulted in the force communicating itself between objects.

Adopting quantum theory, the beams of magnetism imagined by Maunder and his contemporaries became understandable. They were clouds of smashed atoms, each constituent particle carrying a small electrical charge that would strike the Earth's magnetic field and disturb it. Many trillions of such particles could be released by each solar flare, representing tens of billions of tons of matter.

As experimenters became increasingly adept in the manipulation of electrons using quantum theory as their guide, microelectronic technology was born. This led to the computer revolution, which now permeates every pore of scientific inquiry, especially the exploration of space using robotic probes.

With the technology of space probes at their disposal, astronomers have come to see the true extent of the Sun's interaction with the planets and have made great progress in understanding the exotic processes they call *space weather*. Currently they know that the pallid outer atmosphere of the Sun revealed during a total solar eclipse is composed of gas at millions of degrees Celsius. The temperature strips the electrons from their atoms, leaving a seething mass of varying electricity and magnetism that is continually blown out into space in all directions. This outward flow of coronal matter is known as the solar wind and carries the energy that creates the bubble of magnetism surrounding the Solar System.

At times of sunspot minimum, the Sun feeds the solar wind into space at a steady speed in all directions. When sunspot maximum arrives, the solar wind becomes gusty with beams of particles streaming off into space at high speed. Annie Maunder photographed these beams during the 1898 Indian eclipse. It is the strength of the solar wind that causes the daily variation of the compass needle to move in lockstep with the sunspots. At solar maximum, the solar wind leaves the Sun in a turbulent fashion and distorts the Earth's magnetic readings more. At solar minimum, when the solar wind is more uniform, the daily variation quietens, too.

Beyond the continual solar wind, solar flares can cause huge eruptions of coronal particles. Known as a *coronal mass ejection*, such an eruption is almost certainly what many of the 1860 eclipse watchers saw taking place. However, they failed to recognize the importance of the event because they did not know what a "normal" corona looked like. It was only after Warren De la Rue's successful eclipse photograph that reliable records began to be kept and compared. Eventually it was noticed that the corona was always more disturbed around sunspot maximum. Solar flares and their associated coronal mass ejections are the cause of the Earth's magnetic storms.

Thanks to the sophisticated instruments of SOHO and other spacecraft, astronomers can now finally reconstruct the dramatic events surrounding Carrington's flare. As Maunder deduced and Hale measured, a sunspot is just the visible manifestation of a magnetically active region on the Sun. The spot forms when the movement of electrified gas in the

Sun creates a tightly squeezed loop of magnetism that bursts through the solar surface like a pulled thread on a woolen sweater. At the foot of the loop, the magnetism cools the gas, rendering it darker than the surrounding surface gas. The more powerful the magnetic loop, the larger and darker the spot. As Carrington and Maunder watched complex sunspot groups develop, they were actually witnessing the congregation of magnetic loops. The more magnetic loops that burst through the Sun's surface, the more tortured the sunspot group appears. Buffeted by the movement of solar gas in their vicinity, the loops totter thousands of kilometers above the incandescent surface, twisting together until they collapse into a smaller, more stable configuration. When this happens, the energy of a million atomic bombs is unleashed from the magnetic loops, exploding into space as a solar flare.

It takes just eight minutes for the radiation from a solar flare to cross the 93 million miles of space between the Sun and the Earth. Most of the energy is carried in a torrent of X-rays, but in the very largest flares a small portion of the energy can be expelled as visible light. This is what happened on 1 September 1859, taking Carrington by surprise when the brilliant spots of white light appeared above the sunspot. Unseen by him but carrying the majority of the flare's power, X-rays were also striking the Earth. They electrified the particles in the atmosphere, changing the electrical and magnetic properties of Earth's topmost atmospheric layer, which was recorded by magnetic needles at Kew. The timing of Carrington's observation and the Kew readings coincided because the X-rays arrived at exactly the same time as the white light.

This first strike passed in a matter of minutes, allowing the Kew instruments to settle again. It was the lull before the storm. As the solar flare ripped through the Sun's outer atmosphere, it had snared a vast cloud of electrically charged particles in its wake, initiating a coronal mass ejection. Over the course of the next several hours, 10 billion tons of electrons and protons were expelled from the Sun's outer atmosphere and placed on a direct collision course with Earth. Traveling more slowly than the light and X-rays, although still at an extraordinary speed of over 1,500 miles per second, the cloud of particles did not strike Earth until about 17.5 hours later. It was this collision that generated the unprecedented auroras, the magnetic storms, and the surges of current through the telegraph system. Finally, half a day later, the passing cloud left Earth behind.

Coronal mass ejections (CMEs) were definitively identified in the 1970s when space-based telescopes and satellites began continually watching the Sun. At times of sunspot minimum there is perhaps one CME somewhere on the Sun every week. During the maximum activity of the solar cycle, this rises to 2–3 a day. From particularly complex sunspot groups, as proved during the Halloween 2003 storms, the flaring and coronal mass ejections can be almost continuous.

In the first years of the twenty-first century, Dr. Bruce Tsurutani of NASA's Jet Propulsion Laboratory found himself wondering just how powerful the Carrington magnetic storm had been. During the latter decades of the twentieth century, satellites and other space weather equipment had recorded a number of frighteningly large storms. Yet none of them had produced global auroras to the extent reported for the Carrington event. Tsurutani wondered whether Carrington's had been the largest of them all.

All indications suggested that it was colossal, the unprecedented sightings of auroras across the tropical latitudes being particularly telling, but a definitive proof of the storm's power seemed impossible. In his search for answers, Tsurutani found eleven other major storms, but frustratingly, whenever anyone talked about the Carrington event, they invariably used the Kew readings, which had jumped off the scale.

The first huge flare and magnetic storm recorded with modern equipment took place in August 1972 in the declining days of the Apollo Moon landings. The program's penultimate mission had returned to Earth on 27 April, and preparations were underway for the launch of Apollo 17. Solar maximum was over and the sunspot numbers were decreasing, but the Sun had one last surprise of the cycle left. On 4 August, a solar flare triggered a coronal mass ejection that flung tens of billions of tons of solar particles into space. Space-borne instruments recorded the sleet of particles and returned astonishing numbers to Earth. For every hour the storm had raged, space around Earth was engulfed in nine times the yearly radiation limit set for terrestrial radiation workers. The storm had lasted 15.5 hours. If astronauts had been on the surface of the Moon or in flight, they would have received a fatal dose of radiation within the first ten hours of the storm.

The next super flare took place on 13 March 1989. This time it was during the lead-up to solar maximum. The resultant buffeting of the Earth's magnetic field induced such large surges of current along Earth's power lines that controllers of the Hydro-Québec's transmission grid in Canada scrambled to protect their power generation equipment. At 2:44 P.M. the solar storm surged so powerfully that their efforts failed and 6 million people across Québec were blacked out for over nine hours. This was just one of a number of similar power station emergencies that occurred across the North American continent, contributing to a repair bill of some $100 million dollars during the cycle.

During his research into the Carrington event, Tsurutani traveled to India for an academic conference. He was sitting with Professor Gurbax Lakhina of the Indian Institute of Geomagnetism. Over dinner they discussed each other's research, and Tsurutani mentioned his frustration at not being able to find a definitive reading for the Carrington event. The next day, Lakhina produced a leather-bound book from the institute's archives. The book was filled with figures containing the precious data.

The Indian Institute of Geomagnetism had begun life in 1826 as the Colaba Observatory in Bombay. Established by the East India Company to provide astronomical measurements and timekeeping to assist navigation, it had been upgraded by Edward Sabine to become one of his magnetic observatories during the magnetic crusade in the early 1840s. Although funding for the crusade was long over by 1859, locally interested parties had continued the magnetic work. They used equipment decades behind Greenwich and Kew in capability, but operated it with accuracy and diligence. Instead of a continuously recording photographic drum, the readings were taken manually by operators looking through tiny telescopes to observe the twist on the magnets suspended on silken threads many feet away. Lakhina had found the original ledger in which the measurements were recorded.

The ledger revealed that those long-dead operators had routinely taken readings every hour. This was the case in the early hours of 2 September. At 10 A.M., the reading showed that something dramatic was taking place, the magnets having become agitated. The operators stepped up the readings to every fifteen minutes and then to every five minutes as the storm began to rage. Squinting through their telescopes, they followed the storm throughout the rest of the day, through the night, and into the evening of 3 September, increasing the time between readings as things returned to normal. Their readings were transcribed into the very book that Tsurutani was now consulting.

The crucial reading as far as he was concerned was taken around 11:30 A.M. It showed the maximum deflection of the magnet during the storm. Most importantly, it was a clear reading that had not plunged off the scale. Once back at JPL, Tsurutani and his colleagues began plugging the numbers into his computer model and discovered the truth about the Carrington event. The magnetic storm of 1859 was more intense than that of 1989 by some three times and beat the 1972 event as well. The Halloween 2003 magnetic storms had been comparatively mild, falling some five times below the intensity of the Carrington event.

Carrington's had indeed been the perfect solar storm. As Tsurutani's results began to circulate, more and more researchers found their interest in the events of 1859 rekindled. The spectacular auroras drew their attention because many of the reports had mentioned the redness of the light. Eyewitnesses talked of "a thousand fantastic figures, as if painted with fire on a black ground" and of an aurora so bright that it gave out "red light so vivid that the roofs of the houses and leaves of the trees appeared as if covered in blood." The researchers knew that this was a characteristic of truly huge magnetic storms and was triggered by massive quantities of electrons striking the oxygen atoms 125–312 miles high in the atmosphere. Later, as the storm progressed, the electrons pushed their way deeper into the atmosphere extracting green light from the oxygen atoms they hit.

In a storm as large as Carrington's, the heavier protons also have a part to play. Protons are usually deflected by the Earth's magnetic shield, but in powerful storms they are driven into the atmosphere alongside the electrons. Once there, they induce ultraviolet auroras, invisible to human eyes, but which drive chemical reactions. The proton storms induce the formation of nitrates, and these heavy molecules then sink through the atmosphere to ground level, where they become lost in the daily chemical churning of the Earth's surface. A small percentage of the nitrates fall over the Arctic or the Antarctic and suffer a different fate, however. They become locked in suspended animation, sealed in the ice as successive snow layers build up on top of them.

The speed with which the auroras blanketed Earth following Carrington's flare told Dr. Margaret Shea of the Air Force Research Laboratory, Bedford, Massachusetts, that the flare had certainly packed enough of a punch to force protons into Earth's atmosphere. That meant it would have driven nitrate chemistry and locked away the evidence in the ice sheets at the Earth's poles.

Polar scientists have been perfecting ice core extraction and analysis so that they can study the tiny bubbles of Earth's atmosphere locked inside. Because the ice sheets are so deep, the scientists can reach back centuries to monitor climatic conditions such as the buildup of atmospheric pollution during the Industrial Revolution. Shea knew that the nitrates from such ice cores would have been measured as well. Putting together a team of academics from across the United States, she obtained data from an ice core retrieved from Greenland in 1992. The cylindrical core had been as thick as a man's upper arm and as long as a corridor upon its extraction. The ice spanned the time period from 1561 to 1950. It was first cut into manageable lengths, and then each piece was stood upright on a warm plate and melted so that the collected water could be analyzed for nitrates.

Obtaining this data, Shea's team discovered seventy major nitrate deposits, each one coming from a solar flare during the 389 years recorded by the ice core. Using satellite-borne measurements for 1950 onward, they found another eight solar flares. Each one of these seventy-eight events had forced 2 billion protons through every square centimeter of the Earth's atmosphere. To cut the number of events to a manageable data set, the team took the super flare of August 1972 as their reference point. That left them with nineteen super flares that fired more than 5 billion protons through every square centimeter of the Earth's atmosphere. Of these nineteen, one towered four times larger than the benchmark. Calculating the time at which this monster storm had assaulted the Earth, the team pinned it to autumn 1859—it was the Carrington event. A staggering 20 billion protons had sandblasted every square centimeter of our planet's atmosphere during the storm.

During 2004, a joint meeting of the Canadian and American Geophysical Unions convened in Montreal. Attracting hundreds of scientists with a variety of interests, the conference organizers found that, in the light of the Halloween 2003 storms, dozens of them wanted to present their new investigations into the events of 1859. A day and a half of the week-long meeting was turned over to the discussions.

The four factors necessary for a perfect solar storm are (1) the coronal mass ejection caused by the flare must be moving fast, (2) it must be directed straight at Earth, (3) it must be intense rather than a sprawling long-lived storm, and finally (4) the magnetic field carried by the coronal mass ejection must lie in exactly the opposite direction from Earth. Carrington's storm satisfied all of these points, whereas all other storms have failed on at least one count. For example, the Halloween flares did not have the correctly aligned magnetic field to produce as much damage as they otherwise might have done. This realization led the scientists to wonder what would happen if a Carrington-sized event hit Earth today. With our heavy reliance on electrically sensitive technology, Edward Cliver of the Air Force Research Laboratory, Massachusetts, described it as a "worst-case scenario."

If such an event were to engulf Earth today, we should first expect widespread disruption to radio and telephone communications. The electrification of the upper atmosphere would inhibit telecommunications that rely on radio waves, and mobile phones could be affected, too.

Power stations would be placed at severe risk by the huge currents that the magnetic storm induces in the power lines. In unprotected stations these influxes of power would melt transformers, blacking out cities and putting the elderly and infirm at risk, especially if the storm took place in winter. Oil pipelines would carry the currents into fuel stations, risking their sensitive equipment as well.

The toll on satellites would be steep. The charges carried on the electrons and protons would overload and short-circuit the hearts and minds of our technological space voyagers, endangering such systems as GPS navigation. Even if the electricity did not destroy them, the erosion of the solar panels would reduce their available power. The heating of the Earth's atmosphere would cause it to balloon, dragging down the lower satellites from orbit. The March 1989 storm had a measurable effect on over a thousand Earth-orbiting satellites.

The health of astronauts on the International Space Station might hang in the balance, as proved by the extreme radiation doses given out by the August 1972 superflare. Closer to home, airlines would scramble to reroute their flights, pulling them away from the polar regions and reducing their altitude to embed them in the thicker, lower atmosphere. Without these measures, passengers could receive the equivalent of more than ten chest X-rays during a single flight.

This is why spacecraft such as the venerable watchdog SOHO are essential, to monitor and warn us of the impending danger. Apart from simply alerting us when a solar flare is seen to erupt, giving us between fifteen and thirty hours' warning of the incoming storm, scientists using SOHO are beginning to feel they understand enough about the Sun to start issuing some long-range forecasts.

The Sun lapsed into the quiescent phase of its solar cycle in 2006. For twenty-one of February's twenty-eight days, the Sun was spot free. During this time, scientists from America's National Center for Atmospheric Research, based in Boulder, Colorado, used SOHO data to predict that the next solar maximum is going to be the most active in the last fifty years, possibly the last four hundred years. The scientists developed a computer simulation for the way gas moves about inside the Sun. SOHO had revealed vast "conveyor belts" of moving gas that skim matter from near the surface and drag it deep into the Sun's interior. The passage through the interior takes decades, depending on how fast the gas is circulating. This seems to submerge the dying magnetic regions that have been sunspots and slowly rejuvenates them. The magnetic regions then rise to the surface once more, reincarnated as a new generation of sunspots.

With their computer simulation, Dr. Mausumi Dikpati and her team fed in data from the last eighty years and successfully explained the scale of each solar maximum during that time. Their confidence boosted, they ran the program for the forthcoming solar maximum, expected to start sometime between 2010 and 2012. The results predict 30–50 percent more activity than the 2003–2004 maximum—perhaps exactly the conditions needed to finally present us with another Carrington flare.

Beyond the damage such an event might cause to humans and their technology, some scientists are now willing to reconsider the effect that the Sun has on the Earth's weather systems. It seems that William Herschel's ideas about sunspots and wheat prices may not be so absurd after all.

THIRTEEN

The Cloud Chamber

Two hundred years after William Herschel urged the Fellows of the Royal Society to investigate the links between sunspots and Earth's climate, two Israeli scientists found themselves doing just that. Dr. Lev A. Pustilnik of the Israel Cosmic Ray Centre, Tel Aviv, and Dr. Gregory Yom Din, Golan Research Institute, Kazrin, used modern statistical methods to reevaluate Herschel's ideas. At the end of their analysis in 2003, they concluded that the great master of astronomy had been right after all: there does appear to be a link between wheat prices in England during the seventeenth century and solar activity. Wheat prices were higher at solar minimum than at solar maximum, meaning that the crop was more difficult to grow at solar minimum than at solar maximum. This led to a relative shortage of wheat and an inflation of the price.

Although Herschel failed to convince anyone at the time of these connections and received a lampooning for his efforts, Pustilnik and Yom Din showed that it is the master astronomer's third great scientific discovery, standing shoulder to shoulder alongside Uranus and infrared radiation.

Suggestions of a climate link with sunspots have always been hampered by the fact that no one could conceive of a mechanism by which the magnetic activity of the Sun could be transported into the weather-bearing layers of the Earth's atmosphere. The most obvious idea was that the Sun varied its brightness. However, ever since John Herschel and others pioneered devices for measuring the energy of the Sun in the mid-1800s, astronomers have gathered evidence showing that the light from the Sun remains all but constant throughout the solar cycle.

In recent decades, various spacecraft have maintained a perpetual watch from orbit and have clearly measured that the Sun's brightness varies by just 0.1 percent between sunspot maximum and sunspot minimum. Daily and weekly variations of the same magnitude also take place and are linked to the comings and goings of individual sunspots. However, most scientists find it hard to believe that a meager change of just 0.1 percent in the energy output of the Sun could produce any real variation in the Earth's climate. In 1997, however, a plausible mechanism by which the Sun might affect Earth's climate was revealed. It had nothing to do with sunlight and possessed an eerie resonance with Edward Sabine's 1852 announcement that Earthly magnets varied with the solar cycle.

By analyzing records from various weather satellites, covering the period April 1979 to December 1992, Dr. Henrik Svensmark and Dr. Eigil Friis-Christensen of the Danish Meteorology Institute revealed that the Earth's cloud cover varied in lockstep with a phenomenon known to be tied to the solar cycle. They called their discovery "a missing link in solar-climate relationships." The phenomenon at the heart of the link was the influx of cosmic rays. These mysterious rays had been discovered during the early years of the twentieth century. They cascade down to Earth from space and are composed of particles similar to those released by the Sun but carry much more energy. Their exact origin remains unknown, but astronomers suspect they hail from exploding stars spread throughout space, thousands of light-years away, and from the centers of galaxies millions of light-years away. When they strike the upper levels of Earth's atmosphere, they generate showers of other particles that rain down into the lower atmosphere, colliding with the atoms and molecules there.

Physicists have been monitoring cosmic ray activity on Earth since the middle of the twentieth century and have clearly seen that cosmic ray numbers drop at times of high solar activity. They believe this is because the wind of particles from the Sun becomes more powerful at solar maximum, inflating the magnetic field of the Sun and deflecting the incoming cosmic rays. Day-to-day readings confirm this, with cosmic ray strikes dropping noticeably in the aftermath of large solar storms.

Svensmark and Friis-Christensen surprised the scientific community by showing that the fraction of the Earth covered by clouds varied with the number of cosmic rays slicing into the atmosphere. According to their data, the more cosmic rays that hit the Earth, the cloudier the weather becomes. At solar minimum, when cosmic ray strikes are highest, our planet is 3–4 percent cloudier than at solar maximum. Although clouds trap some measure of heat in our atmosphere, this is more than compensated for by the quantity of sunlight they reflect into space. Therefore, a cloudy Earth means a cooler Earth, giving credence to Herschel's assertion that times of fewer sunspots lead to poorer harvests and higher wheat prices.

Scientists might have resurrected their interest in Herschel's ideas sooner if they had taken more seriously one of Walter Maunder's sunspot claims. The date was 1922, and although well into retirement, Maunder felt compelled to work again. He had already been forced back to Greenwich between 1914 and 1918, when the domes were left empty by the call-up of the assistants to fight in the First World War. Now he returned to work for himself. There was a sunspot observation that was glaringly obvious to him, yet all those around him were ignoring it. He had first tried to draw attention to the peculiarity in the 1890s, having been amazed that no one had understood its importance before. Now, aged seventy-one and stricken with a persistent abdominal complaint, Maunder felt compelled to try once more.

He explained that according to the extensive collection of historical sunspot records, first compiled in the mid-1800s by German astronomer Gustav Spörer, it was clear that the dark motes had been rare visitors to the solar surface between the years 1645 and 1715. The sighting of a spot in 1671 had generated widespread excitement among the scientific community for it had been twenty years since a similar occurrence. Astronomers at the time noted that, in the years immediately following Galileo's first use of the telescope to record sunspots, the solar blemishes had been abundant, but in the decades after the mid-1600s there was a marked dearth. The Sun's return to vigor in 1715 gave rise to the great aurora that so surprised the Royal Society that it dispatched Edmond Halley to investigate and report on the phenomenon. Somehow, in the excitement of the aurora, the potential importance of the Sun's seventy years of magnetic calm became lost.

Maunder analyzed the few spots seen between 1645 and 1715 and found that a weak solar cycle could be traced out. In reporting his findings, he wrote that just as in a deeply inundated country, where only the loftiest objects will raise their heads above the flood and enable one to trace out the configuration of the submerged countryside, so the spots seemed to mark out the solar maximums of a "sunken spot curve."

Maunder realized that such variability in the strength of the solar cycle would have profound effects on the magnetic Sun-Earth connection and tried to draw this to people's attention. As had happened during the 1890s with his first attempts to publicize this long-term variability in the solar cycle, no one wanted to know. Maunder's work languished in dusty libraries for half a century until, in the 1970s, Dr. Jack Eddy of the High Altitude Observatory, Colorado, noticed two coincidences: one ironic, the other profound. The first was that the years of the "Maunder Minimum," as Eddy called it, coincided almost perfectly with the reign of Louis XIV of France, *le Roi Soleil.* This was ironic because of the second coincidence: few people felt much of the Sun's heat during those years because the king's reign coincided with the worst years of European weather for over a thousand years. It was a time of extremely harsh winters and became known as the Little Ice Age. In Holland, the canals froze for months on end, and in England frost fairs were organized each year on the solid surface of the River Thames. From a modern perspective, the imagery may appear romantic, but for the millions of people who relied on sunshine for good harvests it was terrible. With so many hovering at subsistence levels, food shortages during the Little Ice Age brought great hardship and suffering.

Eddy's realization that the Little Ice Age coincided with the virtual disappearance of sunspots forced astronomers and climatologists to reconsider the role of the Sun in climate change.

Eddy's interest in astronomical history began when he had been teaching solar physics at the University of Colorado. He found that by recounting historical anecdotes to highlight the way nineteenth-century astronomers had struggled to develop the concepts that were presently daunting the students, he could put the undergraduates at their ease. As Eddy read more into the history of the subject, so he came across the fitful discussions of weather patterns being linked to solar activity. To the solar physics community of the 1970s, Eddy included, the idea was anathema. Then one day, he found himself discussing it with Professor Eugene Parker from the University of Chicago. Parker sagely pointed him in the direction of Maunder's ignored papers. Eddy read them in total disbelief. Convinced that Maunder had been deluded by nothing more than a lack of observations, he determined to search for records that would fill in the gaps and prove the 1645–1715 Maunder Minimum never took place.

Eddy began searching libraries and archives in his spare time. The more manuscripts he found, the more he felt he was "deciphering the Dead Sea scrolls of solar physics" and the more fascinated he became by what he was reading. Each report softened Eddy's initial disbelief. The writings imparted genuine excitement at the opportunity to study a spot, and Eddy began to believe that some great interruption had indeed occurred in the solar cycle. Then disaster: midway through his research, budget cuts descended on the High Altitude Observatory, and Eddy found his name on the list of employees whose services were no longer required.

With a wife and four children to support, he dropped his Maunder Minimum research and began a frantic search for more employment. After a string of rejections, Eddy became desperate. When he was offered a temporary job with NASA, writing a volume for their official history of their space station Skylab, he accepted and soon found that the job presented an unexpected bonus.

He had to travel to a number of universities in order to interview the scientists responsible for the Skylab mission, and this allowed him to use the various libraries at the establishments he was visiting. He began searching for more historical records and recommenced his stalled work on Maunder. Eventually he collected so many sunspot observations from before and after the Maunder Minimum that he realized a genuine drop in the Sun's activity levels was the only explanation for the lack of observations between 1645 and 1715.

To bolster this conclusion, he repeated the painstaking exercise for sightings of auroras and found a staggering tenfold increase in the number of auroral reports following the end of the Maunder Minimum. Still not completely satisfied, he asked himself what lines of investigation were open to him that would not have been available to Maunder.

The answer was the analysis of carbon in tree rings. Every year during the growing season, trees absorb carbon dioxide from the atmosphere and use the carbon atoms to build new cells, widening their trunks. This growth hardens over the winter, creating a tree ring. When cosmic rays hit the atmosphere they can alter the carbon to an isotope known as carbon-14. This combines with oxygen to produce a carbon dioxide and can then be absorbed by trees. Thus, the amount of carbon-14 in any given tree ring betrays that year's cosmic ray intensity.

In the years of the Maunder Minimum, when solar magnetic activity was severely depressed, Eddy reasoned that the cosmic ray flux would be high. This would create a higher proportion of carbon-14 than in years of normal solar activity when the Earth was better shielded by the Sun's magnetism. Looking at the tree-ring data, Eddy found exactly what he was looking for and much more. Besides the clear signature for the Maunder Minimum, there was another similar period stretching from 1460 to 1550, before the invention of the telescope. Eddy called this the Spörer Minimum after the astronomer whose work had inspired Maunder. Intriguingly there was also a long stretch of very low carbon-14 readings, stretching from 1100 to 1250. This seemed to indicate that the Sun had been intensely active at that time, providing the Earth with a highly effective magnetic shield against the cosmic rays. These dates fell within what climatologists referred to as the Medieval Warm Period, when temperate northern latitudes were generally warmer. This period of benign temperatures, drier conditions, and stable seas had allowed the Vikings to colonize Iceland and Greenland. The Greenland colony produced so much wheat that it exported the crop back to Scandinavia. This period was also a time when giant sand dunes roamed the great planes of America because there was too little rain to allow stabilizing grasses to grow in them.

Intrigued by the idea of a solar connection to the Medieval Warm Period, Eddy looked at his sunspot records. Although the years of interest were centuries before the invention of the telescope, he had collected naked-eye sunspot records from the Orient, and sightings of auroras as well. He found an intensification of both naked-eye sunspots and auroras during a two hundred year period centered on 1180.

A clear trend was emerging from the data: fewer sunspots mean lower magnetic activity, which leads to more cosmic ray hits and lower temperatures. Working against considerable skepticism from his scientific peers, Eddy pulled together his lines of research and went public in 1976. He published his findings in America's most prestigious scientific journal, *Science*, figuring that this work was of interest to more scientists than just astronomers. This time, where Spörer and Maunder had failed to arouse interest, Eddy succeeded. His trump card had been to show the coincidence between solar activity and considerable changes in the Earth's climate. He was also helped by the fact that concerns about global warming were beginning to gain momentum. His paper began a debate that continues today as researchers investigate the role of the Sun in past and present climate change.

Since the nineteenth century, global temperatures have risen by an average of 0.6 degrees Celsius in total. The majority of climatologists believe that this is predominantly due to mankind's industrial activities releasing pollution into the atmosphere that traps solar energy, raising our planet's temperature. A much smaller group believes that the Sun's variation is an important factor, perhaps rivaling or even exceeding the human contribution.

At the center of the debate about the Sun's contribution to global warming is the precise nature of Svensmark and Friis-Christensen's link between cosmic rays and cloud cover. This is difficult to resolve because the details of how clouds form is still something of a puzzle. Scientists know that water droplets need something to condense around in order for the cloud to grow. So-called aerosol particles of between 4.1 and 40 millionth of an inch across are good for this. They can be placed in the atmosphere by volcanic activity or the burning of fossil fuels. The question is: Can cosmic rays catalyze the formation of further aerosol particles and so lead to the formation of more clouds?

There is a clue from the pioneering work of the particle physicists who lived at the turn of the twentieth century. In their investigations they discovered that electrically charged particles attract water droplets, forming clouds. They exploited this behavior by building devices called cloud chambers to reveal the otherwise invisible subatomic realms. They filled their cloud chambers with air and water vapor and fired electrically charged particles through the mixture. During their passage, the charged particles collided with the molecules of the air and imparted electrical charges to them. These attracted the water vapor, forming cloudy trails that could be seen and photographed.

With their 1997 study, Svensmark and Friis-Christensen showed that the entire Earth might be a cloud chamber, responding to the bombardment of subatomic particles from deep space. Cosmic ray intensity at Earth varies by 15 percent between solar maximum and solar minimum and, next to the variations in the compass, represents the largest measurable effect of solar activity near the Earth's surface. Just as the protons released in solar eruptions leave fingerprints in the polar ice, so do the cosmic rays. Instead of nitrate molecules, cosmic rays produce an isotope of the element beryllium, known as beryllium-10. When Pustilnik and Yom Dim confirmed Herschel's wheat prices claim, they used beryllium-10 data from ice cores instead of sunspot observations.

Stoking the fires of the Sun's contribution to global warming is that there is growing evidence that the Sun's magnetic activity is reaching an eight thousand-year high. Again, the data come from tree rings. Dr. Sami Solanki of the Max Planck Institut für Sonnensystemforschung, Germany, and collaborators made a grand study of carbon-14 in tree rings and used it to infer the level of solar activity throughout all of recorded human history. According to these results, the past seventy years have seen more solar activity than at any other time in 8,000 years, including the Medieval Warm Period. A separate study by Professor Mike Lockwood and colleagues at Rutherford Appleton Laboratory, Oxfordshire, confirmed the result, suggesting further that the Sun's magnetic activity has more than doubled since 1901. With more magnetic activity, more cosmic rays would be diverted and fewer clouds would form, leading to a warmer Earth. Some see this as compelling circumstantial evidence that global warming is driven by today's solar magnetic activity. Others argue that while the Sun may indeed have an effect, it has now been superseded by human-made pollution. Clearly, some way to test the Sun's effects has to be found.

Unfortunately, the waters are muddied because climate investigations are often politically charged. Some industrial lobbies and governments seize upon any hint of natural warming as a means of avoiding pollution control. On the other hand, environmental pressure groups can sometimes be philosophically opposed to admitting even a small solar effect on climate.

In 2000, a consortium of fifty-six scientists from universities and research establishments across Europe, America, and Russia drew together to plan an experiment to investigate the contribution of cosmic rays to the cloudiness of the Earth. Known as CLOUD, which somewhat humorously stands for Cosmics Leaving OUtdoor Droplets, the experiment will send a beam of high-energy protons through a newly constructed cloud chamber, designed to mimic the properties of the Earth's atmosphere. Detectors will then gauge the reaction of the test atmosphere to the pseudo-cosmic rays. It is currently being constructed, and the team expects to take their first readings during 2008, using the particle accelerator on the French-Swiss border at CERN, the European Organization for Nuclear Research, to provide the proton beam.[1]

If the story of the sun kings has anything to teach us, it is surely that coincidence is often the marker of hidden reality. Indeed, scientists today find themselves in a situation uncannily reminiscent of that experienced by the nineteenth-century astronomers who tried to understand the relevance of sunspots to magnetic storms. Whether the Sun plays a large role in global warming or not, its action in mediating the cosmic rays tells us that the Earth is more closely tied to the wider Universe than even the Victorians realized.

The sixteenth-century poet John Donne wrote famously that no man is an island. Thanks to the work begun by the sun kings of the nineteenth century and continued by those of today, we know that no planet is an island, either. If John Herschel were alive, looking at the balance of evidence, he might well have cause to repeat his words of 150 years ago: "We stand on the verge of a vast cosmical discovery such as nothing hitherto imagined can compare with."

[1] CERN originally stood for Conseil Européen pour la Recherche Nucléaire (European Council for Nuclear Research). In 1954, the name of the organization was changed to Organisation Européenne pour la Recherche Nucléaire (European Organization for Nuclear Research). Faced with an awkward new acronym, OERN, the organization's leaders decided to retain CERN.

EPILOGUE

Magnetar Spring

On 27 December 2004, the largest burst of gamma rays ever recorded cut through the solar system. Smothered in the radiation, satellites instantly began transmitting alert messages to their masters on Earth. As the torrent swept past our planet, part of it bounced off the Moon, and struck Earth again. When astronomers triangulated the blast they found that it came not from the Sun but from deep space. Tracing the blast backward, they found just one celestial object from which it could have originated: the supposed dead heart of a star, just twenty kilometers in diameter and lying some 50,000 light-years away. Known as a magnetar, it is one of a rare breed of celestial objects that contains the most powerful magnetic fields known to exist in nature. If you could magically transport a magnetar to sit halfway between our planet and the Moon, the strength of its magnetic field would wipe every credit card on Earth.

As the astronomers analyzed the gamma-ray data, the numbers became staggering. The magnetar eruption had released more energy into space in a tenth of a second than the Sun shines into space in 100,000 years. The realization that such a distant object could deluge Earth with so much radiation dumbfounded astronomers. They immediately convened a conference to share their data, entitled "A Giant Flare from a Magnetar: Blitzing the Earth from Across the Galaxy."

One of the speakers was Professor Umran Inan of Stanford University in California. He explained that he had been recording the very low frequency radio waves produced by the Earth's uppermost atmosphere at the time of the blast. What his equipment recorded that day astonished him. The gamma rays were far more powerful than anything he had seen the Sun release, and ripped atoms apart across the entire hemisphere of Earth facing the blast. The atmosphere took more than an hour to recover.

Almost 150 years since Carrington saw the first solar swallow of summer, astronomers had been given their first glimpse of magnetar spring.

Bibliography

I have endeavored to list the entries only once and in the chapters where they are most appropriate.

Prologue: The Dog Years

Brekke, Pål (2005) SOHO and solar flares, private communication.

Foullon, C., Crosby, N., and Heynderickx, D. (2005) Towards interplanetary space weather: Strategies for manned missions to mars, *Space Weather* 3, S07004, doi:10.1029/2004SW000134.

Gentley, I. L., Duldig, M. L., Smart, D. F., and Sheas, M. A. (2005) Radiation dose along North American transcontinental flight paths during quiescent and disturbed geomagnetic conditions, *Space Weather* 3, S01004, doi: 10.1029/2004SW000110.

Hildner, Ernest (2005) Space Weather Services at NOAA/SEC: Update. 2nd Symposium on Space Weather, San Diego.

Hogan, Jenny (2004) Sunspot sunset, *New Scientist* 181, no. 2430, 9.

Iles, R.H.A., Jones, J.B.L., and Smith, M. J. (2005) Halloween 2003 Storms: Providing Space Weather Services for Aviation Operations. 2nd Symposium on Space Weather, San Diego.

Jansen, F. (2004) Technical failures or effects due to the space weather storms in the period October/November 2003. Published on http://www.www.uni-greifswald.de/.

Joint USAF/NOAA Report of Solar and Geophysical Activity (2003) SDF number 302.

Jones, Bryn (2005) Space Weather–Operational and Business Impacts. Airline Space Weather Workshop Report, Boulder.

Jones, Bryn, Iles, R.H.A., and Smith, M. J. (2005) Integrating Space Weather Information into Global Aviation Operations. 2nd Symposium on Space Weather, San Diego.

Kappenman, John G. (2005) Impacts to Electric Power Grid Infrastructures from the Violent Sun-Earth Connection Events of October–November 2003. 2nd Symposium on Space Weather, San Diego.

Murtagh, William J. (2005) Redefining the Solar Cycle: An Operational Perspective. 2nd Symposium on Space Weather, San Diego.

NOAA Extreme Solar Flare Alert (2003) Space Weather Advisory Bulletin 03-5.

NOAA Intense Active Regions Emerge on the Sun (2003) Space Weather Advisory Bulletin 03-2.

NOAA Solar Active Region Produces Intense Solar Flare (2003) Space Weather Advisory Bulletin 03-3.

NOAA Space Weather Outlook (2003) Space Weather Advisory Outlook 03-44.

NOAA Space Weather Outlook (2003) Space Weather Advisory Outlook 03-47.

NOAA Space Weather Scales, www.sec.noaa.gov/NOAAscales/.

Tsurutani, B. T., Judge, D. L., Guarnieri, F. L., Gangopadhyay, P., Jones, A. R., Nuttall, J., Zambon, G. A., Didkovsky, L., Mannucci, A. J., Iijima, B., Meier, R. R., Immel, T. J., Woods, T. N., Prasad, S., Floyd, L., Huba, J., Solomon, S. C., Straus, P., and Viereck, R. (2005) The October 28, 2003 extreme EUV solar flare and resultant extreme ionospheric effects: Comparison to other Halloween events and the Bastille Day event, *Geophys. Res. Lett.* 32, no. 3, L03S09.

Unknown, SOHO Web pages: www.esa.int/science/soho, sohowww.estec.esa.nl/, soho.esa.int/science-e/www/area/index.cfm?fareaid=14.

Chapter 1: The First Swallow of Summer

Bruzelius, Lars (2005) Clipper ships and aurora, private communication.

Carlowicz, Michael J., and Lopez, Ramon E. (2002) *Storms from the Sun.* Joseph Henry Press, Washington, DC.

Carrington, R. C. (1860) Description of a singular appearance seen in the Sun on September 1, 1859. *Monthly Notices of the Royal Astronomical Society* 20: 13.

Davis, T. N. (1982) Carrington's solar flare. Alaska Science Forum (www.gi.alaska.edu/ScienceForum), Article 518.

Hodgson, R. (1860) On a curious appearance seen in the Sun. *Monthly Notices of the Royal Astronomical Society* 20: 15.

______ (1861) On the brilliant eruption on the Sun's surface, 1st September 1859. Report of the 13th Meeting of the BAAS, held at Oxford 1860, 36. John Murray, London.

Loomis, Elias (1860) The great auroral exhibition of Aug. 28th to Sept. 4th, 1859, and the geographical distribution of auroras and thunder storms—5th Article. *American Journal of Science and Arts* (2nd series) 30, no. 88: 79.

______ (1860) The great auroral exhibition of Aug. 28th to Sept. 4th, 1859—6th Article. *American Journal of Science and Arts* (2nd series) 30, no. 90: 339.

______ (1860) The great auroral exhibition of August 28th to September 4th, 1859—2d Article. *American Journal of Science and Arts* (2nd series) 29, no. 85: 92.

______ (1860) The great auroral exhibition of August 28th to September 4th, 1859—3d Article. *American Journal of Science and Arts* (2nd series) 29, no. 86: 249.

______ (1860) The great auroral exhibition of August 28th to September 4th, 1859—4th Article. *American Journal of Science and Arts* (2nd series) 29, no. 87: 386.

______ (1861) The great auroral exhibition of Aug. 28th to Sept. 4th, 1859—7th Article. *American Journal of Science and Arts* (2nd series) 32, no. 94: 71.

______ (1861) The great auroral exhibition of Aug. 28th to Sept. 4th, 1859, and on auroras generally—8th Article. *American Journal of Science and Arts* (2nd series) 32, no. 96: 318.

Stewart, Balfour (1861) On the great magnetic disturbance which extended from August 28 to September 7, 1859, as recorded by photography at the Kew Observatory. *Phil. Trans.* 151: 423.

Unknown (1851) The new clipper ship *Southern Cross*, of Boston. *The Boston Daily Atlas*, May 5 edition.

______ (1860) The great auroral exhibition of August 28th to September 4th, 1859. *American Journal of Science and Arts* (2nd series) 28, no. 84: 385.

Chapter Two: Herschel's Grand Absurdity

Gribbin, John (2005) *The Fellowship: The Story of a Revolution*. Penguin, London.

Hall, Marie Boas (2002) *All Scientists Now: The Royal Society in the Nineteenth Century*. Cambridge University Press, Cambridge.

Herschel, William (1796) On the method of observing the changes that happen to the fixed stars; with some remarks on the stability of the light of our Sun. *Phil. Trans.*, 166.

——— (1800) Experiments on the refrangibility of the invisible rays of the Sun. *Phil. Trans.* 90: 284.

——— (1800) Investigation of the powers of the prismatic colours to heat and illuminate objects; with remarks, that prove the different refrangibility of radiant heat. To which is added, an inquiry into the method of viewing the Sun advantageously, with telescopes of large apertures and high magnefying powers. *Phil. Trans.* 90: 255.

——— (1800) On the nature and construction of the sun and fixed stars. *Phil. Trans.* 85: 46.

——— (1801) Observations tending to investigate the nature of the Sun, in order to find the causes of symptoms of its variable emission of light and heat; with remarks on the use that may possibly be drawn from solar observations. *Phil. Trans.* 91: 265.

Hoskin, Michael (2003) *The Herschel Partnership: As Viewed by Caroline.* Science History Publications, Cambridge, U.K.

Hoskin, Michael (ed.) (2003) *Caroline Herschel's Autobiography*. Science History Publications, Cambridge, U.K.

Hoskin, Michael (2005) Unfinished business: William Herschel's sweeps for nebulae. *History of Science* 43.

——— (2005) William Herschel's sweeps for nebulae. *The Speculum* 4, no. 1: 38.

Hufbauer, Karl (1991) *Exploring the Sun: Solar Science since Galileo.* Johns Hopkins University Press, Baltimore.

Lovell, D. J. (1868) Herschel's dilemma in the interpretation of thermal radiation. *Isis* 59, no. 1: 46.

Lubbock, C. (1933) *The Herschel Chronicle.* Cambridge University Press, Cambridge.

Schaffer, Simon (1980) "The Great Laboratories of the Universe": William Herschel on matter theory and planetary life. *Journal for the History of Astronomy* 11: 81.

——— (1980) Herschel in Bedlam: Natural history and stellar astronomy. *British Journal for the History of Science* 13, no. 45: 211.

——— (1981) Uranus and the establishment of Herschel's astronomy. *Journal for the History of Astronomy* 12: 11.

Soon, Willie, and Baliunas, Sallie (2003) *The Varying Sun and Climate Change.* Fraser Forum, Vancouver.

Taylor, R. J. (ed.) (1987) *History of the RAS.* 2 volumes. Blackwell Scientific Publications, Oxford.

Unknown. Royal Society Web site: www.royalsoc.ac.uk.

______ Somerset House Web site: www.somerset-house.org.uk.

Chapter Three: The Magnetic Crusade

Blöckh, Alberto (1972) *Consequences of Uncontrolled Human Activities in the Valencia Lake Basin in The Careless Technology: Ecology and International Development.* Natural History Press, New York.

Cawood, John (1979) The magnetic crusade: Science and politics in early Victorian Britain. *Isis* 70, no. 254: 493.

Cliver, E. W. (1994) Solar activity and geomagnetic storms: The first 40 years. *Eos, Transactions, American Geophysical Union* 75, no. 49: 569, 574–575.

Gilbert, William (translated by Sylvanus P. Thompson, 1900) *On the Magnet.* London.

Good, Gregory (2004) *On the Verge of a New Science: Meteorology in John Herschel's Terrestrial Physics, from Beaufort to Bjerknes and Beyond: Critical Perspectives on the History of Meteorology.* International Commission on History of Meteorology, Weilheim, Germany.

Hawksworth, Hallan, and Atkinson, Francis B. (1926) *A Year in the Wonderland of Trees.* Charles Scribner's Sons, New York.

Helferich, Gerard (2004) *Humboldt's Cosmos.* Gotham Books, New York.

Hoskin, Michael (1993) *Bode's Law and the Discovery of Ceres*, 21–33. Astrophysics and Space Science Library 183: Physics of Solar and Stellar Coronae, J. Linski and S. Serio (eds.). Dordrecht, Kluwer.

Kollerstrom, N. (1992) The hollow world of Edmond Halley. *J. History of Astronomy* 23: 185.

Malin, S.R.C. (1996) Geomagnetism at the Royal Observatory, Greenwich. *Quart. J. Roy. Astron. Soc.* 372: 65.

Malin, S.R.C., and Barraclough, D. R. (1991) Humboldt and the earth's magnetic field. *Quart. J. Roy. Astron. Soc.* 32: 279.

Millman, Peter M. (1980) The Herschel dynasty—Part II: John Herschel. *J. Roy. Astron. Soc. Can.* 74, no. 4: 203.

Pumfrey, Stephen (2002) *Latitude and the Magnetic Earth.* Icon Books, London.

Reingold, Nathan (1975) Edward Sabine, in *Dictionary of Scientific Biography*, vol. 12, p. 49. Charles Scribner's Sons, New York.

Robinson, P. R. (1982) Geomagnetic observatories in the British Isles. *Vistas in Astronomy* 26: 347.

Stern, David P. (2002) a millennium of geomagnetism. *Reviews of Geophysics* 40, no. 3: 1-1–1-30. (Also available on-line: www.phy6.org/earthmag/mill_1.htm.)

Weigl, Engelhard (2001) Alexander von Humboldt and the beginning of the environmental movement. *International Review for Humboldtian Studies, HiN* 2, no. 2.

Chapter Four: The Solar Lockstep

Buttman, Günther (1970) *The Shadow of the Telescope: A Biography of John Herschel.* Charles Scribner's Sons, New York.

Carrington, R. C. (1851) An account of the late total eclipse of the Sun on July 28, 1851, as observed at Lilla Edet. *Pamphlets of the Royal Astronomical Society* 42, no. 9.

——— (1851) On the longitude of the observatory of Durham, as found by chronometric comparisons in the year 1851. *Monthly Notices of the Royal Astronomical Society* 12: 34.

——— (1851) Solar eclipse of July 28, 1851, Lilla Edet, on the Göta River. *Monthly Notices of the Royal Astronomical Society* 12: 55.

Chapman, Allan (1996) *The Victorian Amateur Astronomer: Independent Astronomical Research in Britain, 1820–1920.* Wiley-Praxis, Chichester, U.K.

Forbes, Eric G. (1975) Richard Christopher Carrington, in *Dictionary of Scientific Biography* vol. 3, p. 92. Charles Scribner's Sons, New York.

Herschel, John (1852) Letter to Edward Sabine 15/3/52. Royal Society Sabine Archives.

——— (1852) Letter to Michael Faraday 10/11/52. Royal Society Herschel Archives.

Keer, Norman C. (2000) *The Life and Times of Richard Christopher Carrington B.A. F.R.S. F.R.A.S. (1826–1875).* Privately published.

Kollerstrom, Nick (2001) Neptune's discovery: The British case for co-discovery. http://www.ucl.ac.uk/sts/nk/neptune/.

Lindop, Norman (1993) Richard Christopher Carrington (1826–1875) and solar physics. Project Report for M.Sc. Astronomy and Aeronautics, University of Hertfordshire, U.K.

Meadows, A. J., and Kennedy, J. E. (1982) The origin of solar-terrestrial studies. *Vistas in Astronomy* 25: 419.

Rochester, G. D. (1980) The history of astronomy in the University of Durham from 1835 to 1939. *Quart. J. Roy. Astron. Soc.* 21: 369.

Sabine, Edward (1852) Letter to John Herschel 16/3/52. Royal Society Herschel Archives.

Schwabe, Heinrich (1843) Solar observations during 1843. *Astronomische Nachrichten* 20, no. 495: 234.

Scott, Robert Henry (1885) The history of the Kew Observatory. *Proceedings of the Royal Society of London* 39: 37.

Standage, Tom (2000) *The Neptune File*. Penguin, London.

Unknown (1876) Richard Carrington obituary. *Monthly Notices of the Royal Astronomical Society* 36: 137.

Chapter Five: The Day and Night Observatory

Carrington, R. C. (1855) Letter to G. B. Airy. Cambridge University Library, RGO Archive 6/235, 618–620.

______ (1857) *A Catalogue of 3735 Circumpolar Stars observed at Redhill, in the years 1854, 1855, and 1856, and reduced to mean positions for 1855*. Eyre and Spottiswoode, London.

______ (1857) Notice of his solar-spot observations. *Monthly Notices of the Royal Astronomical Society* 17: 53.

______ (1858) On the distribution of the solar spots in latitude since the beginning of the year 1854. *Monthly Notices of the Royal Astronomical Society* 19: 1.

______ (1858) On the evidence which the observed motions of the solar spots offer for the existence of an atmosphere surrounding the Sun. *Monthly Notices of the Royal Astronomical Society* 18: 169.

Cliver, Edward W. (2005) Carrington, Schwabe, and the Gold Medal. *Eos, Transactions, American Geophysical Union* 86, no. 43: 413, 418.

Lightman, Bernard (ed.) (1997) *Victorian Science in Context*. University of Chicago Press, Chicago.

Schwabe, H. (1856) Extract of a letter from M. Schwabe to Mr. Carrington. *Monthly Notices of the Royal Astronomical Society* 17: 241.

Unknown (1856) Summary of Richard Carrington's recent tour of European observatories. *Monthly Notices of the Royal Astronomical Society* 17: 43.

Chapter Six: The Perfect Solar Storm

Burley, Jeffery, and Plenderleith, Kristina (eds.) (2005) *A History of the Radcliffe Observatory Oxford—The Biography of a Building*. Green College, Oxford.

Helfferich, Carla (1989) The rare red aurora. Alaska Science Forum (www.gi.alaska.edu/ScienceForum), Article 918.

Loomis, Elias (1869) The Aurora Borealis or Polar Light. *Harper's New Monthly Magazine* 39, No. 229.

Marsh, Benjamin V. (1861) The aurora, viewed as an electric discharge between the magnetic poles of the Earth, modified by the Earth's magnetism. *American Journal of Science and Arts* (2nd series) 31, no. 93: 311.

Newton, H. A. (1895) Biographical memoir of Elias Loomis, in *Biographical Memoirs*, vol. 3, p. 213. National Academy of Sciences, Washington, DC.

Odenwald, Sten. www.solarstorms.org.

Siegel, Daniel M. (1975) Balfour Stewart, in *Dictionary of Scientific Biography*, vol. 13, p. 51. Charles Scribner's Sons, New York.

Walker, Charles V. (1861) On the magnetic storms and earth-currents. *Phil. Trans.* 151: 89.

Chapter Seven: In the Grip of the Sun

Clerke, Agnes M. (1902) *A Popular History of Astronomy during the Nineteenth Century*. 4th edition. A. and C. Black, London.

Farber, Eduard (ed.) (1966) Bunsen's methodological legacy, in *Milestones of Modern Chemistry*, p. 15. Basic Books, New York.

Kirchhoff, G. R. (1861) On a new proposition in the theory of heat. *Phil. Mag.* 21, Series 4: 241.

______ (1861) On the chemical analysis of the solar atmosphere. *Phil. Mag.* 21, Series 4: 185.

Meadows, A. J. (1984) The origins of astrophysics, in *The General History of Astronomy*, vol. 4A (ed. Owen Gingerich). Cambridge University Press, Cambridge.

Meadows, Jack (1970) *Early Solar Physics*. Pergamon Press, London.

Porter, R. (ed.) (1994) Joseph von Fraunhofer, in *The Biographical Dictionary of Scientists*. Oxford University Press, Oxford.

Rosenfeld, L. (1973) Gustav Kirchhoff, in *Dictionary of Scientific Biography*, vol. 17, p. 379. Charles Scribner's Sons, New York.

Schacher, Susan G. (1970) Robert Bunsen, in *Dictionary of Scientific Biography*, vol. 2, p. 586. Charles Scribner's Sons, New York.

Watson, Fred (2005) *Stargazer: The Life and Times of the Telescope*. Allen and Unwin, Melbourne.

Chapter Eight: The Greatest Prize of All

Airy, G. B. (1860) Letter to Richard Carrington. RGO Archive 6/146, 58-9.

______ (1860) Account of observations of the total solar eclipse of 1860, July 18, made at Hereña, near Miranda de Ebro; with a notice of the general proceedings of "The Himalaya Expedition for Observation of the Total Solar Eclipse." *Monthly Notices of the Royal Astronomical Society* 21: 1.

Barnes, Melene (1973) Richard C. Carrington. *Journal of the British Astronomical Association* 83, no. 2: 122.

Carrington, R. C. (1858) Information and suggestions to persons who may be able to place themselves within the shadow of the total eclipse of the Sun on 7th September, 1858. Royal Astronomical Society Pamphlets, vol. 42.

______ (1859) Letter to John Herschel 13/3/59. Royal Society Herschel Archives.

______ (1860) An eye-piece for the solar eclipse. *Monthly Notices of the Royal Astronomical Society* 20: 189.

______ (1860) Formulae for the reduction of Pastorf's observations of the solar spots. *Monthly Notices of the Royal Astronomical Society* 20, 191.

______ (1860) Letter to George Airy. RGO Archive 6/146, 56-7.

______ (1860) Letter to John Herschel 2/5/60. Royal Society Herschel Archives.

______ (1860) On some previous observations of supposed planeatary bodies in transit over the Sun. *Monthly Notices of the Royal Astronomical Society* 20: 192.

______ (1860) Proposed new design for vertically placed divided circles. *Monthly Notices of the Royal Astronomical Society* 20: 190.

______ (1861) Letters to the Vice Chancellor and Senate of Cambridge University. Syndicate Papers in Cambridge University Senate Archives.

de la Rue, Warren (1862) The Bakerian Lecture: On the total solar eclipse of July 18th, 1860, observed at Rivabellosa, Near Miranda de Ebro, in Spain. *Phil. Trans.* 152: 333.

Eddy, J. A. (1974) A nineteenth-century coronal transient. *Astron. & Astrophys.* vol. 34: 235.

Faye, M. (1860) Total solar eclipse of July 18, 1860. *American Journal of Science and Arts* (2nd series) 29, no. 85: 136.

Hingley, Peter D. (2001) The first photographic eclipse? *Astronomy and Geophysics* 42: 1.18.

Hoyt, Douglas V., and Schatten, Kenneth H. (1995) A revised listing of the number of sunspot groups made by Pastorff, 1819 to 1833. *Solar Physics* 160, no. 2: 393.

Unknown (1861) Auctioneer's catalogue of sale of Redhill property. Royal Astronomical Society Pamphlets, vol. 42.

Chapter Nine: Death at the Devil's Jumps

Airy, George Biddell (1868) Comparison of magnetic disturbances recorded by the self-registering magnetometers at the Royal Observatory, Greenwich, with magnetic disturbances deduced from the corresponding terrestrial galvanic currents recorded by the self-registering galvanometers of the Royal Observatory. *Phil. Trans.* 158: 465.

——— (1870) Note on an extension of the comparison of magnetic disturbances with magnetic effects inferred from the observed terrestrial galvanic currents; and discussion of the magnetic effects inferred from the galvanic currents on days of tranquil magnetism. *Phil. Trans.* 160: 215.

——— (1872) On the supposed periodicity in the elements of terrestrial magnetism, with a period of 261/3 days. *Proceedings of the Royal Society of London* 20: 308.

Carrington, R. C. (1863) *Observations of the Spots on the Sun, from November 9th 1853 to March 24th 1861, Made at Redhill.* Williams and Norgate, London.

——— (1863) On the financial state and progress of the Royal Astronomical Society, Royal Astronomical Society Pamphlets, vol. 42.

——— (1865) Revenue account versus cash account—A Breeze. Royal Astronomical Society Pamphlets, vol. 42.

——— (1866) Appeal on the accounts at a special meeting of the Royal Astronomical Society. Royal Astronomical Society Pamphlets, vol. 42.

Ellis, William (1906) Sun-spots and magnetism—A retrospect. *The Observatory* 29, no. 376: 405.

Lanzerotti, Loius J., and Gregori, Giovanni. P (1986) *Telluric Currents: The Natural Environment and Interactions with Man-Made Systems: The Earth's Electrical Environment.* The National Academies Press, Washington, DC.

Stewart, Balfour (1864) Remarks on sun spots. *Proceedings of the Royal Society of London* 13: 168.

Unknown (1871) Murderous assault, *The Hampshire Chronicle*, August 26, p. 7.

——— (1871) The Farnham tragedy. *The Hampshire Chronicle*, September 9, p. 8.

——— (1871) The tragedy near Farnham. *The Hampshire Chronicle*, September 2, p. 8.

——— (1872) The tragedy at the Devil's Jumps, Farnham. *The Surrey Advertiser*, March 30, p. 2.

______ (1875) *The Surrey Advertiser*, December 11.
______ (1875) Inquests, *The Times*, December 7, p. 5.
______ (1875) Inquests, *The Times*, November 22, p. 5.
Young, C. A. (1896) *The Sun* (Appleton, New York).

Chapter Ten: The Sun's Librarian

Airy, George Biddell (1874) *Testimony before the Devonshire Commission, Royal Commission on Scientific Instruction and the Advancement of Science, Minutes of Evidence, Appendices, and Analyses of Evidence*, vol. 2. Eyre and Spottiswoode, London.

Becker, Barbara J. (1993) Eclecticism, opportunism, and the evolution of a new research agenda: William and Margaret Huggins and the origins of astrophysics. Ph.D. diss., Johns Hopkins University, Baltimore. Available online: http://eee.uci.edu/clients/bjbecker/huggins/.

Chapman, Allan. George Biddell Airy, F.R.S. (1801–1892): A centenary commemoration. *Notes and Records of the Royal Society of London* 46, no. 1 (1992): 103.

Forbes, E. G., Meadows, A. J., and Howse, H. D. (1975) *Greenwich Observatory: The Royal Observatory at Greenwich and Herstmonceux, 1675–1975*, volumes 1–3. Taylor and Francis, London.

Jevons, W. S. (1878) Commercial crises and sun-spots. *Nature* 19: 33.

______ (1882) The solar commercial cycle. *Nature* 26: 226.

______ (1875) Influence of the sun-spot period on the price of corn. *Nature* 16.

Kinder, Anthony John (2006) Edward Walter Maunder, FRAS (1851–1928). Part I—His Life & Times. In preparation.

Maunder, Annie S. D., and Maunder, E. Walter (1908) *The Heavens and Their Story*. Epworth Press, London.

Maunder, E. Walter (1900) *The Royal Greenwich Observatory: A glance at Its History and Work*. Religious Tract Society, London.

Peart, Sandra (2000) "Facts Carefully Marshalled," in the *Empirical Studies of William Stanley Jevons*, vol. 33, p. 352 of *History of Political Economy*. Duke University Press, Durham, NC.

Porter, Theodore M. (1986) *The Rise of Statistical Thinking*. Princeton University Press, Princeton, NJ.

Soon, Willie Wei-Hock, and Yaskell, Steven H. (2004) *The Maunder Minimum and the Variable Sun-Earth Connection*. World Scientific Publishing, Singapore.

Stewart, Balfour (1885) Note on a preliminary comparison between the dates of cyclonic storms in Great Britain and those of magnetic disturbances at the Kew Observatory. *Proceedings of the Royal Society of London* 38: 174.

Strange, Alexander (1872) On the insufficiency of existing national observatories. *Monthly Notices of the Royal Astronomical Society* 32: 238.

——— (1874) *Testimony before the Devonshire Commission, Royal Commission on Scientific Instruction and the Advancement of Science, Minutes of Evidence, Appendices, and Analyses of Evidence*, volume 2. Eyre and Spottiswoode, London.

Unknown (1882) The light in the sky. *New York Times*, April 18.

White, Michael (2000) Some difficulties with sunspots and Mr. Macleod: Adding to the bibliography of W. S. Jevons. *History of Economics Review* 31.

Chapter Eleven: New Flare, New Storm, New Understanding

Buchwald, Jed Z. (1976) Sir William Thomson (Baron Kelvin of Largs), in *Dictionary of Scientific Biography*, vol. 13, p. 374. Charles Scribner's Sons, New York.

Cliver, E. W. (1994) Solar activity and geomagnetic storms: The corpuscular hypothesis. *Eos, Transactions, American Geophysical Union* 75, no. 52: 609, 612–613.

——— (1995) Solar activity and geomagnetic storms: From M regions and flares to coronal holes and CMEs. *Eos, Transactions, American Geophysical Union* 76, no. 8: 75, 83.

Cortie, A. L. (1903) Sun-spots and terrestrial magnetism. *The Observatory* 26, no. 334: 318.

Ellis, William (1880) On the Relation between the diurnal range of magnetic declination and horizontal force, as observed at the Royal Observatory, Greenwich, during the years 1841 to 1877, and the period of solar spot frequency. *Phil. Trans.* 171: 541.

——— (1892) On the simultaneity of magnetic variations at different places on occasions of magnetic disturbances, and on the relation between magnetic and earth current phenomena. *Proceedings of the Royal Society of London* 52: 191.

——— (1904) The auroras and magnetic disturbance. *Monthly Notices of the Royal Astronomical Society* 64: 228.

Hale, George Ellery (1892) A remarkable solar disturbance. *Astron. Astrophys.* 11: 611.

______ (1908) On the probable existence of a magnetic field in sun-spots. *Astrophysical Journal* 28: 315.

______ (1931) The spectrohelioscope and its work, Part III: Solar eruptions and their apparent terrestrial effects. *Astrophysical Journal* 73: 379.

Kellehar, Florence M. (1997) George Ellery Hale, Yerkes Observatory Virtual Museum: astro.uchicago.edu/yerkes/virtualmuseum/.

Maunder, E. Walter (1892) Note on the history of the great sun-spot of 1892 February. *Monthly Notices of the Royal Astronomical Society* 52: 484.

______ (1899) *The Indian Eclipse 1898: Report of the Expeditions Organized by the British Astronomical Association to Observe the Total Solar Eclipse of 1898, January 22*. Hazell, Watson, and Viney Ltd., London.

______ (1904) Further note on the "great" magnetic storms, 1875–1903, and their association with sun-spots. *Monthly Notices of the Royal Society* 64: 222.

______ (1904) The "great" magnetic storms, 1875 to 1903, and their association with sun-spots, as recorded at the Royal Observatory, Greenwich. *Monthly Notices of the Royal Society* 64: 205.

______ (1905) The solar origin of terrestrial magnetic disturbances. *Popular Astronomy* 13, no. 2: 59.

______ (1906) The solar origin of terrestrial magnetic disturbances. *Journal of the British Astronomical Society* 26: 140.

______ (1907) Abstract of lecture delivered before the Association at the meeting held on December 19 on Greenwich sun-spot observations and some of their results. *Journal of the British Astronomical Society* 27: 125.

Pang, Alex Soo Jung-Kim (2002) *Empire and the Sun: Victorian Solar Eclipse Expeditions*. Stanford University Press, Stanford, CA.

Proctor, Richard A. (1891) *Other Suns than Ours*. W. H. Allen and Co., London.

Thomson, Sir William (Lord Kelvin) (1892) Presidential address on the anniversary of the Royal Society. *Nature* 47, no. 1205: 106.

Unknown (1892) Brilliant electric sight: A wonderful exhibition of northern lights. *New York Times*, February 14.

______ (1904) Meeting of the Royal Astronomical Society, Friday 1904 January 8. *The Observatory* 27, no. 341: 75.

______ (1904) Meeting of the Royal Astronomical Society, Friday 1904 November 11. *The Observatory* 27, no. 351: 423.

______ (1905) Meeting of the British Astronomical Association, Wednesday 1905 February 22. *The Observatory* 28, no. 356: 170.

______ (1905) Meeting of the Royal Astronomical Society, Friday 1905 March 10. *The Observatory* 28, no. 356: 157.

______ (1905) Meeting of the Royal Astronomical Society, Friday 1905 January 13. *The Observatory* 28, no. 354: 77.

——— (1907) Death of Lord Kelvin. *The Times*, 18 December.

Warner, Deborah Jean (1974) Edward Walter Maunder, in *Dictionary of Scientific Biography*, vol. 9, p. 183. Charles Scribner's Sons, New York.

Chapter Twelve: The Waiting Game

Dikpati, Mausumi, de Toma, Giulana, and Gilman, Peter A. (2006) Predicting the strength of solar cycle 24 using a flux-transport dynamo-based tool. *Geophys. Res. Lett.* 33: L05102.

Lovett, Richard A (2004) Dark side of the sun. *New Scientist*, 4 September, p. 44.

Odenwald, Sten (1999) Solar storms. *Washington Post*, 10 March.

Shea M. A., Smart, D. F., McCracken, K. G., Dreschhoff, G.A.M., and Spence, H. E. (2004) Solar proton events for 450 Years: The Carrington event in perspective. *Eos, Transactions, American Geophysical Union* 85, no. 17, Jt. Assem. Suppl. Abstract SH51B-04.

Tsurutani, B. T., Gonzalez, W. D., Lakhina, G. S., and Alex, S. (2003) The extreme magnetic storm of 1–2 September 1859. *J. Geophys. Res.* 108 (A7): 1268, doi:10.1029/2002JA009504.

Unknown (2004) Spacecraft fleet tracks blast wave through solar system. NASA Release 04-217.

Various (2004) Solar and Heliospheric Physics, Session SH43A and SH51B at 2004, Joint Assembly of the AGU.

Wilson, John W., Cucinotta, Francis A., Jones, T. D., and Chang, C. K. (1997) Astronaut protection from solar event of August 4, 1972. NASA Technical Paper 3643.

Chapter Thirteen: The Cloud Chamber

Baliunas, Sallie (1999) Why so hot? Don't blame man, blame the sun. *Wall Street Journal*, August 5.

Beer, J., Tobias, S. M., and Weiss, N. O. (1998) An active sun throughout the Maunder Minimum. *Solar Physics* 181: 237.

Bingham, Robert (2006) The CLOUD proposal, private communication.

Chapman, Allan (1994) Edmond Halley's use of historical evidence in the advancement of science. *Notes and Records of the Royal Society of London* 48, no. 2: 167.

Eddy, Jack (1977) The case of the missing sunspots. *Scientific American*, May, p. 80.

______ (1976) The Maunder Minimum. *Science* 192, no. 4245: 1189.

______ (1980) Climate and the role of the sun. *Journal of Interdisciplinary History* 10, no. 4: 725.

Fastrup, B., Pedersen, E., and 54 others (2000) A study of the link between cosmic ray and clouds with a cloud chamber at the CERN PS. CERN/SPSC 2000-021 SPSC/P317.

Feldman, Theodore S. (year unknown) Solar variability and climate change: A historical overview. Available online at: http://www.agu.org/history/SV.shtml.

Halley, Edmond (1715) An account of the late surprizing appearance of the lights seen in the air, on the sixth of March last. *Phil. Trans.* 29: 406.

______ (1719) An account of the phaenomena of a very extraordinary aurora borealis, seen at London on November 10, 1719. *Phil. Trans.* 30: 1099.

Maunder, E. Walter (1890) Professor Spoerer's researches on sun-spots. *Monthly Notices of the Royal Astronomical Society* 50: 251.

______ (1922) The prolonged sunspot minimum 1645–1715. *Journal of the British Astronomical Society* 32: 140.

McKee, Maggie (2004) Sunspots more active than for 8000 years. New Scientist.com, posted 27 October.

Pustilnik, Lev A., and Yom Din, Gregory (2003) Influence of solar activity on state of wheat market in medieval England. *Proceedings of International Cosmic Ray Conference 2003.* Available online at xxx.lanl.gov/abs/astro-ph/0312244.

Solanki, S. K., Usoskin, I. G., Kromer, B., Schüssler, M., and Beer, J. (2004) Unusual activity on the Sun during recent decades compared to the previous 11,000 years. *Nature* 431: 1084.

Svensmark, Henrik, and Friis-Christensen, Eigil (1997) Variation of cosmic ray flux and global cloud coverage—a missing link in solar climate relationships. *Journal of Atmospheric and Solar-Terrestrial Physics* 59, no. 11: 1225.

Tinsley, Brian A. (2005) Evidence for space weather affecting tropospheric weather and climate. 2nd Symposium on Space Weather, San Diego.

Weart, Spencer (1999) Interview with Jack Eddy, April 21, 1991. Available online at http://www.agu.org/history/sv/solar/index.shtml.

Epilogue: Magnetar Spring

Inan, U., Lehitnen, N., Moore, R., Hurley, K., Boggs, S., Smith, D., and Fishman, G. J. (2005) Massive disturbance of the daytime lower ionosphere by the giant X-ray flare from Magnetar SGR 1806-20. Abstract IAGA2005-A-00844. Available at www.cosis.net.

Soloman, Robert C. (2003) "Magnetars," soft gamma ray repeaters and very strong magnetic fields. Published online at solomon.as.utexas.edu/~duncan/magnetar.html.

Index

actinometer, 60
Adams, John Couch, 61–62; and the Cambridge Observatory, 112–113
airglow, 30
Airy, George Biddell, 67–68, 117–118, 131–133, 151–152; and the 1860 solar eclipse, 100–101; retirement of, 140; and the workforce at Greenwich, 137
American Astronomical Society, 151
Apollo moon landings, 173
Arrhenius, Svante August, 162
atomic components, 170; electrons, 158; protons, 175–176
aurora, 4–5; of 28 August 1859, 84–87; of 2 September 1859, 14–18, 20, 87–89; arc, 15–16; color, 15; corona, 15–16; mapping of by Elias Loomis, 89–91; patch, 14; ray, 15–16; southern aurora, 10; ultraviolet, 175
airline rerouted flights, 6
asteroids. *See* Ceres; Juno; Pallas
Astronomical Society, 44

BAA. *See* British Astronomical Association
BAAS. *See* British Association for the Advancement of Science
Baily, Francis, 101
Baily Beads, 101
Baron Kelvin of Largs. *See* Thompson, William
berylium-10, 186
Blandford, H. F., 140
Bode, Johann Elert, 39
British Association for the Advancement of Science (BAAS), 54–57; acquisition of Kew Observatory in 1842, 68
British Astronomical Association (BAA), 145; expedition to observe the 1898 eclipse in India, 155–158
British Empire, 53
Brougham, Henry, 37–38
Bunsen burner, 95
Bunsen, Robert, and Fraunhofer lines, 95–96

carbon-14, 183–184, 186
Carrington flare, 14, 81–82, 126, 171; modern-day analysis of, 173–178
Carrington, Richard Christopher, 11–24, 60–66, 80–82; and the Cambridge Observatory, 112–115; Churt residence of, 120; death of Rosa, 126–127; "differential rotation" of the Sun, 78; and the Durham Observatory, 62–63, 65–66; and the Furze Hill Observa-

Carrington, Richard Christopher (*con't*) tory, 71–72; and Heinrich Schwabe, 75–77; marriage to Rosa Helen Rodway, 120–125; and the Middle Devil's Jump Observatory, 120; modern-day analysis of Carrington's flare, 173–178; northern star catalog of, 74, 77; and the Radcliffe Observatory, 80–81, 98–99; RAS Gold Medal award in 1859, 76; sale of in 1861, 115–116; and the solar eclipse of 1851, 64–65; and the solar eclipse of 1860, 101–102; solar flare studies of, 14, 81–82, 126, 171; solar observations of, 12–14, 64–65, 73–78; suicide of, 127; sunspot catalog of, 119
Cassini spacecraft, 6; and Halloween flares, 169
cathode rays, 158
Ceres (asteroid), 39–40
CERN, 187
Challis, James, 61–62, 111–112
Christie, William, 140
CLOUD. *See* Cosmics Leaving OUtdoor Droplets
cloud chamber, 185–186
cloud cover and cosmic rays, 180–182
cloud formation, 185–186
CME. *See* coronal mass ejection
computer revolution, 170–171
Comte, Auguste, 95
corona. *See* solar corona
coronal mass ejection (CME), 171–173
cosmic rays; and cloud cover, 180–182, 185–186; proxy data from tree ring analysis, 183–184, 186; and solar activity, 180
Cosmics Leaving OUtdoor Droplets (CLOUD), 186–187

De la Rue, Warren, 99–100, 151; eclipse expedition to Spain in 1860, 100–109; and Kew Observatory photoheliograph, 99–100
deforestation, 48
Devonshire Commission, 131, 133, 135
"differential rotation" of the Sun, 78
diffraction grating, 94
Durham's longitude measurement, 63
Durham Observatory, 62–63, 65–66

Earth's magnetic cloak, 169
eclipse. *See* solar eclipse
economic cycles and 11-year sunspot cycle, 139
Eddy, Jack, 182–185; and the Maunder Minimum, 182–184; and the Spörer Minimum, 184
electrons, 158
Ellis, William, 168; classification of magnetic storms, 160

Famine Commission, 140
FitzGerald, George Francis, 165
flame test, 94–95
Fox Talbot, William and Fraunhofer lines, 94–95
Fraunhofer, Joseph von, 93–94
Fraunhofer lines, 94–97, 126; Earth's atmospheric effects on, 97
Furze Hill Observatory, 71–72; sale of in 1861, 115–116

Galileo, Galilei, sunspot observations of, 28–29
Galileo spacecraft, 30
Galle, Johann Gottfried, 61
Gauss, Carl Friedrich, 53, 55
geographical pole, 48–49
Gilbert, William, 48
glass making, 93–94
global warming, 185–186
Greenwich Observatory, 67–68, 117–118, 133; Edward Walter Maunder's work at, 136–146; receipt of the

Kew photoheliograph, 135; solar photography and spectroscopy at, 136, 140
Gulliver's Travels, 38

Hale, George Ellery, 148; founder of the American Astronomical Society, 151; and the Kenwood observatory, 149; marriage to Evelina, 148; and the Mount Wilson Observatory, 166; and the solar flare of July 1892, 149–150; and the spectroheliograph instrument, 148; and the University of Chicago
Halley, Edmond, 49
Halloween flares, 2–3, 5, 118, 168–169; effect of on Mars, 169; effect of on outer planets of the Solar System, 169
Herschel, John, 41–46, 57–60; actinometer measurements of, 60; cataloging of the southern skies, 55–56; death of, 130; and Fraunhofer lines, 94–95; and terrestrial magnetism, 53–54
Herschel, William, 25–46; and asteroids Ceres, Juno, and Pallas, 39–40; climate change–solar connection proposal, 33; and colored glass eyepieces, 33; and colored light analysis, 34–35; discovery of Uranus by, 26; health of, 42; and infrared rays, 35–36; son, John Herschel, 41–46; sunspot and solar observations of, 28–31, 36–37; sunspot–wheat-prices paper by, 37, 41, 179
Hiorter, O. P., 18
Hodgson, R. Esq., 21–23
Huggins, William, 135–136
Humboldt, Alexander von, 47–49; magnetic studies of, 50–53
Hydro-Québec power disruption, 174

ice core data; analysis of berylium-10 from, 186; analysis of nitrate events from, 176–176
ignis fatuus, 15
Indian Institute of Geomagnetism, 174
Indian Meteorological Department, 138
Indian monsoon failures, 138
infrared radiation, Herschel's observations of, 35–36
International Space Station, 4

Jevons, William Stanley, 139
Juno (asteroid), 40

Kelvin of Largs. *See* Thompson, William
Kenwood observatory, 149
Kew Observatory, 14, 18, 68, 82–83; BAAS acquisition of in 1842, 68; magnetic instrumentation of, 20; and the photoheliograph, 99–100, 106, 134–135, 141–142; Royal Society takeover in 1872, 134; and solar photography, 75
Kirchhoff, Gustav: and Fraunhofer lines, 95–97; laws of, 97; and limelight, 96; and spectral analysis, 96–97

Lake Valencia, Venezuela, 47
Larmor, Joseph, 170; electricity flow theory of, 159, 164–165
Le Verrier, Urbain, 61
limelight, 96
Little Ice Age, 182
loadstone, 48
Loomis, Elias, 83; mapping auroral events of 28 Aug/2 Sept 1859, 89–90; and tornado wind speeds, 83

magnetar, 188–189
magnetic compass, 48–49
magnetic declination, 49
magnetic equator, 48, 50
magnetic inclination, 49–50
magnetic pole, 48–49
magnetic storm, 50, 82–83, 152–154; of 28 August 1859, 83–87; of 2 September 1859, 87–89; classification by William Ellis, 160; correlation with sunspots, 69, 159–162

Mars Odyssey spacecraft and the Halloween flares, 169
Maunder, Edward Walter, 129–130; 1892 sunspot observations of, 146; death of Edith, 145; eclipse expedition to India (1898), 155–158; founder of the British Astronomical Association, 145; and the Greenwich Observatory, 136–146; Lick Observatory invitation, 145; marriage to Annie Scott Dill Russell, 154–155; marriage to Edith Hannah Bustin, 138; photograph of the 1898 eclipse, 157; presentation to RAS fellows, 161–165; sunspot and magnetic storm analysis of, 159–162, 181–182
Maunder Minimum, 182–184
Maxwell, James Clerk, theory of electricity, 153
Medieval Warm Period, 184
Mercury, 52
Middle Devil's Jump Observatory, 120
Midori 2 weather satellite, 4
Mount Wilson Observatory, 166

Napoleonic Wars, 35
Neptune, 61
Newton, Isaac, 29
nitrate formation by proton storms, 175–176
northern star catalog of Carrington, 74, 77

Olbers, Wilhelm Matthäus, 39

Pallas (asteroid), 39–40
periodogram analysis, 163
photoheliograph, 99–100, 106, 134, 141–142; British Empire, 143. *See also* solar photography
photosphere, 97
Piazzi, Giuseppe, 38–39
power grid disruption, 174
Prescott, George, B., 88–89
Prime Meridian, 63
protons, 175–176
proton storms and nitrate formation, 175–176

quantum theory, 170

Radcliffe Observatory, 80–81, 98–99
RAS. *See* Royal Astronomical Society
Royal Astronomical Society (RAS), 22–23, 54; and women members, 144–145. *See also* Astronomical Society
Royal Society, 27–28, 54
Russell, Annie Scott Dill, 145–146; marriage to Edward Walter Maunder, 154–155; photograph of the 1898 eclipse, 157

Sabine, Edward, 54–56, 66–67, 151; magnetic observations of, 68–70; retirement of, 131; solar photography of, 75
Saint Elmo's fire, 10
Schröter, Johann Hieronymus, 32, 40
Schuster, Arthur, 163–164
Schwabe, Heinrich, 51–52; RAS Gold Medal award in 1857, 76; and Richard Christopher Carrington, 75–77; and sunspot cycles, 68–69, 73; sunspot observational methods, 76
Secchi, Pietro Angelo, 111
Simmonds, George Harvey, 72; and Carrington northern star catalog, 77
solar atmosphere, 97
solar corona, 109
solar cycle, 68–69, 73
solar eclipse; of 1851, 64–65; of 1860, 100–116; of 1898, 155–158
solar flare; 4 August 1972 flare, 173; 13 March 1989 flare, 174; Carrington,

14, 81–82, 126, 171, 173–176; Hales 149–150; Halloween flares, 2–3, 5, 118, 168–169
solar photography, 74; at the Kew Observatory, 75. *See also* photoheliograph
solar prominence, 65; viewed during the 1860 solar eclipse, 107–108
solar rotational period of 27 days, 160–161
SOHO. *See* Solar and Heliospheric Observatory
Solar and Heliospheric Observatory (SOHO), 1–7, 171, 178
solar wind, 171
Southern Cross clipper ship, 9–11
space weather, 171
spectral analysis, 96–97
spectroheliograph instrument, 148
Spörer, Gustav, 79
Spörer Minimum, 184
star trackers, 4
Stewart, Balfour, 82–83, 151
Strange, Alexander, 131–136; proposed national astrophysical laboratory of, 131–132, 135
Sun: brightness variations of, 179–180; magnetic field extent of, 169–170
sunspot: 1892 observations by Edward Walter Maunder, 146–148; Carrington sunspot catalog, 119; climate link to, 138–140; correlation with magnetic activity, 69, 159–162, 184; cycle observations by Heinrich Schwabe, 68—69, 73; and economic cycles, 139; formation of, 171–172; Herschel's observations of, 28–31, 36–37

telegraph system: disabling of in 1859, 21–22; 28 August magnetic storm and, 84; 2 September magnetic storm and, 88–89
Tempest, The, 9–10
Thompson, Joseph John and electrons, 158
Thompson, William, a.k.a Baron Kelvin of Largs, 152–154, 159
Titius–Bode law, 39
Titius, Johann Daniel, 39
tree ring analysis for carbon-14, 183–184, 186
Tsurutani, Bruce, and modern-day research on Carrington's flare, 173–175

ultraviolet auroras, 175
Ulysses spacecraft and Halloween flares, 169
Uranus, 61; discovery of, 26

Venus, 30
Voyager 2 spacecraft and Halloween flares, 169

Wesleyan Society, 130
will-o'-the-wisp. See *ignis fatuus*
Wilson, Alexander, 30
Wilson, Patrick, 31
Wolf, Johann Rudolph, 73; sunspot observations of, 73
Wollaston, William, 93

X-rays, 172

Yerkes, Charles, 151
Yerkes Observatory, 151, 166